D0794039

Approaches to Automotive Emissions Control

Approaches to Automotive Emissions Control

Richard W. Hurn, *Editor*

A symposium co-sponsored by
the Division of Fuel Chemistry
and the Division of Petroleum
Chemistry at the 167th Meeting
of the American Chemical
Society, Los Angeles, Calif.,
April 1–2, 1974.

ACS SYMPOSIUM SERIES 1

AMERICAN CHEMICAL SOCIETY

WASHINGTON, D. C.   1974

Library of Congress CIP Data

Approaches to automotive emissions control.
(ACS symposium series; 1)

Includes bibliographical references and index.

   1. Automobile exhaust gas—Congresses. 2. Motor vehicles—Pollution control devices—Congresses.
   I. Hurn, R. W., 1919- ed. II. American Chemical
Society. III. American Chemical Society. ACS symposium
series; 1.

TD886.5.A66   629.2'53   74-22443
ISBN 0-8412-0212-5   ACSMC8 1 1-207 (1974)

ACS Symposium Series

Robert F. Gould, *Series Editor*

FOREWORD

The ACS Symposium Series was founded in 1974 to provide a medium for publishing symposia quickly in book form. The format of the Series parallels that of its predecessor, Advances in Chemistry Series, except that in order to save time the papers are not typeset but are reproduced as they are submitted by the authors in camera-ready form. As a further means of saving time, the papers are not edited or reviewed except by the symposium chairman, who becomes editor of the book. Papers published in the ACS Symposium Series are original contributions not published elsewhere in whole or major part and include reports of research as well as reviews since symposia may embrace both types of presentation.

CONTENTS

Preface ix

1. The Influence of Fuel Composition on Total Energy Resources 1
   G. P. Hinds, Jr. and W. A. Bailey, Jr.

2. Impact of Automotive Trends and Emissions Regulations on Gasoline
   Demand 19
   D. H. Clewell and W. J. Koehl

3. Gaseous Motor Fuels—Current and Future Status 43
   W. R. Finger, D. S. Gray, W. J. Koehl, P. E. Mizelle, A. V. Mrstik,
   S. S. Sorem, and J. F. Wagner

4. Fuel Volatility as an Adjunct to Auto Emission Control 60
   R. W. Hurn, B. H. Eccleston, and D. B. Eccleston

5. Pre-engine Converter 69
   N. Y. Chen and S. J. Lucki

6. Low Emissions Combustion Engines for Motor Vehicles 78
   Henry K. Newhall

7. Alternative Automotive Emission Control Systems 99
   E. N. Cantwell, E. S. Jacobs, and J. M. Pierrard

8. The Application of the High Speed Diesel Engine as a Light Duty
   Power Plant in Europe 159
   C. J. Hind

9. Automotive Engines for the 1980's 172
   R. W. Richardson

Index 203

PREFACE

The course of automotive emissions control has profound influence both on attaining improved environmental quality and on meeting the energy needs of the transport sector. Thus this symposium, "Approaches to Automotive Emissions Control," simultaneously addresses two issues that will dominate the national research and development effort for a long time.

The symposium was conceived as a vehicle to foster recognition and critical examination of the interaction between auto emissions control and energy requirements. Taken together, the papers provide excellent background for insight into the interplay of automotive emissions control, fuel economy, and overall energy requirement. These elements of transportation technology have been considered independently for too long. Now, because of the potentially critical problems in meeting fuel demand, a reasonable balance must be drawn between control requirements and the preservation of potential for improved fuel economy.

RICHARD W. HURN

Bartlesville, Okla.
September 9, 1974

The Influence of Fuel Composition on Total Energy Resources

G. P. HINDS, JR. and W. A. BAILEY, JR.

Shell Development Co., Deer Park, Tex. 77536

A meaningful discussion of the subject of this paper is only possible within predefined limits. If all energy used is in the form of electricity generated by solar power or nuclear fusion, fuel composition is of no significance. However, since chemical fuels, particularly liquid chemical fuels, represent a convenient and inexpensive way of storing large amounts of energy in a small, light container, it is probable that they will be used for self-propelled vehicles at least for many years to come. The air pollution problems associated with their use are well on the way to being solved. The chemical fuels can be easily distributed and handled and they can be synthesized from available raw materials. This discussion will be limited to chemical fuels.

Since the combustion products of fuels for oxidation processes must be capable of being handled by the biosphere without damaging it, the useful elemental compositions are limited. Elements whose oxidation products are irritating or toxic (for example sulfur) cannot be considered nor can those whose scarcity precludes their general use. Thus chemical fuels must probably be compounds of carbon, hydrogen, oxygen, and perhaps nitrogen, although inclusion of nitrogen compounds under some circumstances creates problems.

Obviously, during the twentieth century the most widely used liquid fuels have been hydrocarbons, and these compounds have many advantages as chemical fuels. Their physical properties vary widely, making possible the tailoring of fuels to meet varying combustion requirements. Since they have been manufactured in quantity, the technology for their efficient distribution and use has been developed. Most importantly, their energy content per unit mass is higher than that of other eligible compounds. Figure 1 shows the net heat of combustion in Btu/# as a function of molecular weight for hydrocarbons, and some oxygen, nitrogen and sulfur compounds. As would be expected, as the oxygen content of a molecule increases the heat of combustion decreases.

For hydrocarbons, the net heat of combustion is essentially a function of hydrogen content, Figure 2. The curve of Figure 2 is defined by the equation:

$$H_c = 14,100 + 402.5 \times \%w\ H - 4.26\ (\%w\ H)^2 \qquad (1)$$

This equation can be generalized to give an approximation of the heat of combustion of other organic compounds of interest. The generalized equation is:

$$\text{Net } H_c = \left(1 - \frac{\% S}{100} - 1.126 \frac{\% O}{100} - \frac{\% N}{100}\right)\left[14,100 + 402.5\right.$$
$$\left(\frac{(\% H - \frac{\% O}{8})\ 100}{\% C + \% H - \frac{\% O}{8}}\right) - 4.26\left(\frac{(\% H - \frac{\% O}{8})\ 100}{\% C + \% H - \frac{\% O}{8}}\right)^2 \left.\right] + 82 \times \% S \quad (2)$$

Although the equation is based on compounds containing only one kind of hetero atom, it can be used to estimate the heat of combustion of kerogen, shale oils, coal, etc., from the elemental composition. Figure 3 illustrates the agreement between measured and calculated heats of combustion for some hydrocarbons and non-hydrocarbon compounds. Points representing two coals are also shown.

As has been pointed out previously (1), many other properties of hydrocarbons, specific gravity, refractive index, smoke point, etc., can be related to the hydrogen content. Even a property as sensitive to molecular structure as viscosity correlates in a general way with this parameter - consider the difference between graphite and methane. This trend can be illustrated qualitively by the viscosity on the C_{26} saturated hydrocarbons synthesized by API Research Project 42 at Pennsylvania State University (2) - Figure 4. Here the absolute viscosity in centipoise at 210° F is plotted against the number of rings in the molecule. The average value and range of data are indicated. Essentially the same curve can be drawn for aromatic rings, so that hydrogen content alone does not determine viscosity, but the trend of increasing viscosity with decreasing hydrogen content is obvious. (Also certain specific highly sterically hindered structures have been synthesized. The inclusion of data on these structures increases the scatter of the points, but does not affect the general conclusion.)

The manufacture of chemical fuels from crude oil, i.e., petroleum refining, can be considered as the process of adjusting the molecular weight and hydrogen content of the hydrocarbons present to meet product specifications, and the amount of high valued products which can be manufactured from a given crude is limited by the amount of hydrogen available. Nearly all crude oils contain less hydrogen than the most desirable product mix. Typical data for crudes and currently acceptable fuels are shown in Table 1. Only residual fuel oil and asphalt contain less

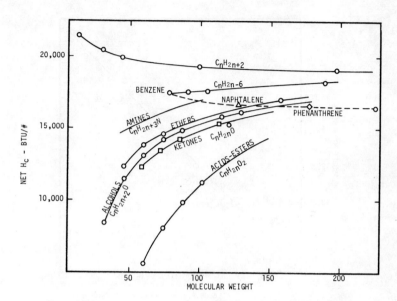

Figure 1. Heat of combustion as a function of molecular weight

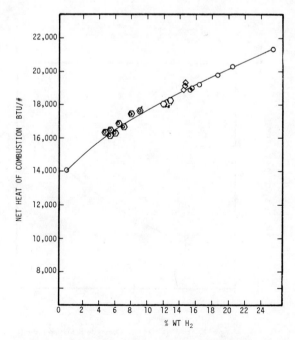

Figure 2. Heat of combustion as a function of hydrogen content

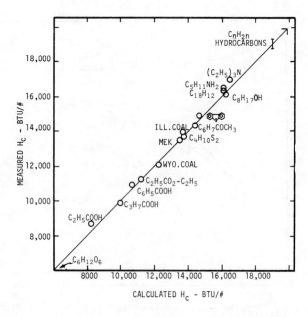

Figure 3. *Calculated and measured heats of combustion*

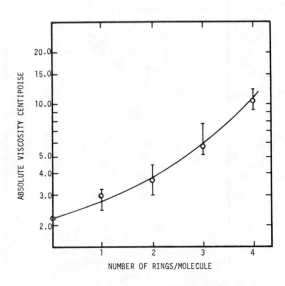

American Petroleum Institute
Figure 4. *Viscosity of saturated pure hydrocarbons
with 26 carbon atoms/molecule*

Table 1

Hydrogen Contents of Fossil Fuels and Some
Current Specification Products

	%w H	
	Typical	(Range)
Crude Oils	12.3	(10-14)
Residues from Crudes	11.8	(9.5-12.5)
Natural Gas	22.5	
Liquified Petroleum Gases	17.5	
"Regular" Gasoline	14.3	
"Premium" Gasoline	13.7	
Aircraft Turbine Fuel	13.8	
Diesel Fuel - No. 2 Furnace Oil	12.3	
Residual Fuel - No. 6 Fuel	10.0	
Asphalt	11.2	

hydrogen than the residue from which many products must be made.
When fuels are derived from other sources, shale, tar sands, or
coal, the problem is aggravated. Not only are these materials
poorer in hydrogen than most crude oils, but they also contain
higher concentrations of sulfur and nitrogen. Compounds of
these elements must be removed to comply with air pollution
regulations, and at the present state of knowledge, their removal
can be accomplished practically only by hydrogenation. Unless
the chemical fuels of the future can be of high molecular weight,
high viscosity, and of poor burning quality, it seems probable
that their production will require the generation of hydrogen.

The usable hydrogen content of the raw material from which
a fuel is to be made has a significant effect on the energy
required to produce it. The energy required to deliver by
pipeline to a Gulf Coast refinery a pound of hydrogen combined
with carbon in crude oil is approximately 300-400 Btu. To
generate the same pound of hydrogen by steam-methane reforming
would require 85,000 Btu. When the starting material contains
sufficient hydrogen so that auxiliary generation is not needed,
the efficiency of energy resource use can be high. The importance to current refining practice of efficient use of hydrogen
is obvious.

When fuels are to be manufactured from raw materials of
lower hydrogen content than that desired in the finished product, additional hydrogen must be generated, probably from water
and excess carbonaceous material. The production of pure hydrogen from coke is a commercially practical, if expensive, process.
Such a process might achieve a thermal efficiency, defined as
the net heat of combustion of the hydrogen produced divided by
the net heating value of the coke needed for the reaction and
to provide the power required for the oxygen plant, separation
equipment, etc., of about 55%. (Current efficiencies are in
the range of 46-47%.) This would be equivalent to the consumption of 6.7 pounds of coke for each pound of hydrogen produced. Using this efficiency, the weight of fuel of a given
hydrogen content which can be produced from a unit weight of
coke can be calculated by simple stoichiometry. (The heat of
formation of hydrocarbons is sufficiently small so that it can
be neglected in this context.) Figure 5 shows this relationship.
The assumptions used in this and subsequent calculations are
summarized in Table 2. By combining this curve with that
relating heat of combustion to hydrogen content, Figure 2, a
curve relating fuel hydrogen content to the fractional reduction
in total available energy can be constructed for the synthesis
of hydrocarbon fuels from coke, air and water, Figure 6. From
this curve it is obvious that unaltered carbon should be used
as a source of energy whenever this is feasible without serious
efficiency penalty. The generation of methane from coke carries
with it a penalty of about 38% of the energy reserves.

The preceding picture is greatly oversimplified. Coals, not
coke, are usually available, and these contain sulfur, which must

Table 2

Assumptions Used in Energy Calculations

Thermal Efficiency of Hydrogen Generation:

Gasification of coal or coke to - 55%
hydrogen only

Steam-Hydrocarbon Reforming   62%

To Remove S, N, or O by Hydrogenation

Hydrogen required for S - 4 moles H_2/mole S -
Thiophene desulfurization

Hydrogen required for N - 4 1/2 moles H_2/mole N - N
removal - from Pyridine and Pyrrole

Hydrogen required for O - 2 1/2 moles H_2/mole O - O
removal - from Esters, Furans, Phenols

The additional H not associated with H_2S, NH_3 and H_2O is
incorporated in the fuel.

Hydrogen in coal or shale oil available to react with impurities.

All char-carbon gasified before additional coal used.

Net Heat of Combustion of fuels:

$$
H_c = \left(\frac{\% \text{ C}}{100} + \frac{\% \text{ H}}{100} - \frac{\% \text{ O}}{800}\right) \left[14,100 + 402.5 \times \left(\frac{(\% \text{ H} - \frac{\% \text{ O}}{8}) \ 100}{\% \text{ C} + \% \text{ H} - \frac{\% \text{ O}}{8}}\right) \right.
$$

$$
\left. - 4.26 \times \left(\frac{(\% \text{ H} - \frac{\% \text{ O}}{8}) \ 100}{\% \text{ C} + \% \text{ H} - \frac{\% \text{ O}}{8}}\right)^2 \right] + 82 \times \% \text{ S}
$$

All % are weight

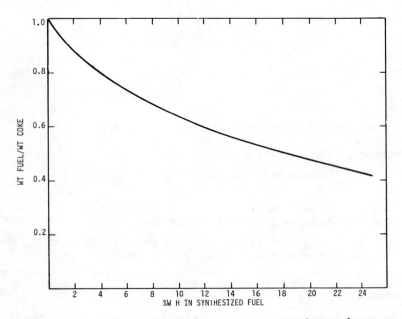

Figure 5.   Weight recovery of hydrocarbons produced from coke

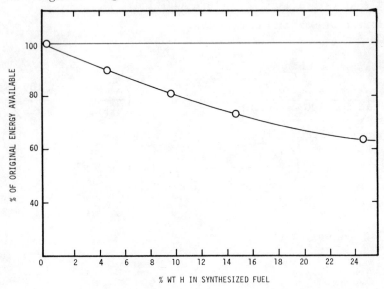

Figure 6.   Energy recovery in hydrocarbons produced from coke

be removed before or after combustion, as well as oxygen and
nitrogen. Shale oil contains more hydrogen than coal, but is
also rich in sulfur and nitrogen compounds which require hydrogen for their elimination. Practical conversion processes
produce a range of compounds of varying hydrogen contents rather
than a single, most desirable one. Literature data on coal hydrogenation in pilot plant equipment show products ranging from
H_2S, NH_3 and CH_4, to residual fuel (3). The presence of impurities, and the limitations of reaction mechanisms, qualitatively
increase the hydrogen requirements to produce fuels of given
properties, and thus reduce the fraction of energy recoverable.

Analyses of the organic material in oil shale (4) and of
the oil derived therefrom by retorting (5), have been published,
as have elemental compositions of several coals (3). Table 3
shows typical data. In the case of shale, it is not unreasonable
to assume that the heat needed for retorting can be supplied by
combustion of the unrecovered fraction of the kerogen, and thus
that the initial raw material is the retorted oil. By making
the assumptions shown in Table 2, it is possible to calculate
the fuel yield and the fraction of the original energy recovered
from shale oil as a function of hydrogen content of the final
fuel. Figures 7 and 8 show these relations.

Similar calculations can be made for coals of varying composition, and curves for an Illinois No. 6 bituminous and a Wyoming
subbituminous are included in Figures 7 and 8. The curves are
based on dry, ash-free coals, and represent an idealized situation. It has been assumed that the oxygen, sulfur and nitrogen
in the coal can be removed by reaction with a portion of the
hydrogen present, and that the remaining hydrogen can be included
in the hydrocarbon fuel manufactured. Only excess carbon is
gasified (with a thermal efficiency of 55%) to produce hydrogen.
The curves in Figures 7 and 8 extend between a point equivalent
to the hydrogen content of the oxygen, sulfur and nitrogen-free
raw material, and the composition of methane. The actual raw
material points (at 100% recovery) are also shown. Obviously,
the higher the usable hydrogen content of the starting material,
the higher will be the percent of the original energy recovered.

Because of the relatively high costs associated with the
processing schemes discussed, it is logical to assume that only
fuels of high value, synthetic natural gas, LPG, gasoline, and
light fuel oils should be produced by these costly synthetic
routes. Since these fuels are the most probable ones for small
self-propelled vehicles, the remainder of this discussion will
consider only the manufacture of them.

Several studies have been made over the years which have
attempted to evaluate octane number of gasoline which minimizes
the cost of transportation to the motorist. In each case the
approach has been similar. First it is necessary to establish
a relationship between compression ratio and mileage at a
constant level of performance; second, to relate octane number

Table 3

Elemental Composition	%w C	%w H	%w N	%w S	%w O
Kerogen – Mahogany Zone Green River Formation	80.5	10.3	2.4	1.0	5.8
Shale Oil – NTU Retort	84.25	11.41	2.02	0.72	1.61
Illinois No. 6 Coal – Dry (ash free basis)	79.50	5.53	1.02	3.54	10.41
Wyoming Subbituminous Coal – Dry (ash free basis)	75.46	4.71	1.10	0.66	18.07

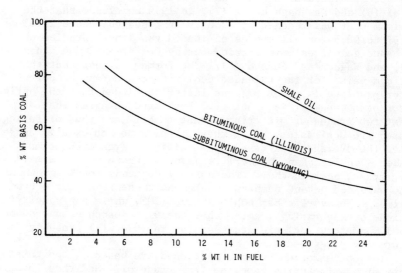

Figure 7. Weight recovery of hydrocarbons synthesized from shale and coal

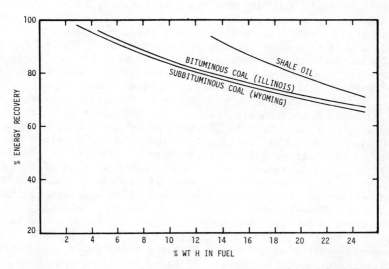

Figure 8. Energy recovery in hydrocarbons synthesized from shale and coal

requirement to compression ratio for the average car on the road;
and third, to estimate the relation between gasoline octane
number and cost. From these data, the gasoline cost per mile
can be calculated as a function of octane number. Duckworth,
et al (6) and Kavanagh et al (7), in 1959 concluded that the
minimum cost was reached at an average research octane level
of 97 to 98.5 for all grades of leaded gasoline. Studies on
unleaded gasolines were carried out by Corner and Cunningham
(8), and Wagner and Russum (9). The former suggest that the
optimum value for the unleaded pool research octane number of
motor gasoline is about 97; the latter concluded that the optimum
motor octane number is about 87. Both studies agree that at
least two grades of differing octane number (and thus differing
costs) are desirable. While these studies do not exactly agree,
since the average pool gasoline sensitivity (research octane
number - motor octane number) is about 8, there is not a major
discrepancy between them. There is by no means total agreement
that unleaded octane numbers as high as these represent a true
optimum. However, other publications (10), while not specifying
optimum octane ratings, agree that the 91 research octane number
level mandatory for unleaded gasoline in 1974 will probably
eventually be exceeded.

The optimization studies mentioned are based on the total
costs of manufacturing gasoline from crude oil, and thus involve
charges for capital and operating costs, as well as raw materials. Thus they may not reflect exactly an optimum fuel composition to maximize total energy resources. However, since the
most practical route to increasing the octane quality of motor
gasolines is to increase the aromatic content, which results in
a decrease in total hydrogen content, it appears that the manufacture of light, low octane number fuels does not represent
a desirable direction for energy conservation. (Ellis has
calculated that, even in the absence of catalytic mufflers, an
increase in gasoline aromatics does not cause a proportionate
increase in reactive exhaust hydrocarbons (10).)

It is not chemically reasonable to postulate a process based
on a reaction so specific that it will produce only a single
hydrocarbon from a variety of starting materials. Some data
are available in the literature on the products obtained by
coal hydrogenation, and the properties of the various products
(3). Table 4 is based on data on the hydrogenation of bituminous coal, and shows yields, properties, and estimated hydrogen content for the various fractions. (The hydrogen contents
and heats of combustions have been estimated from limited physical property data.) From these data and the assumptions in
Table 2, weight and energy recoveries can be calculated. The
reported product distribution contains heavy fuels which do not
meet current specifications. If these fractions are freed of
sulfur and nitrogen, and upgraded to current ATF and No. 2
(diesel) gas oil quality, their hydrogen content must be

Table 4

Data from Hellwig, et al

Component	Basis: 100# Coal Charge				API Gravity	Est. %w H	Assumed Minimum %w H	Add'l H Needed	Yield %w
	Yield %w	# C	# H	# H Added					
$C_1 - C_3$	10.2	8.06	2.14	1.59	---	21.0	21	0	10.2
C_4 - 400°F	18.5	15.72	2.78	1.72	49.2°	15.0	15	0	18.5
400°-680°F	27.5	24.37	3.13	1.49	21.1°	11.4	13.8	0.56	26.8
680°-975°F	12.7	11.57	1.13	0.35	0.3°	8.9	12.3	0.41	12.5
975°F +	12.3	11.56	0.74	-0.04	-20.0°	6.0	12.3	0.80	12.5
Char-Carbon	10.7	10.70	0.0	-0.72	---	0	---	---	10.7
NH_3 H_2S H_2O	14.0	---	1.51	1.51	---	---	---	0.24	16.7
Total	105.9	81.98	11.43	5.9	---	12.2	14.5	2.0	107.9

H Yield from Gasification of Char = 1.61#

increased, and additional hydrogen must be generated. For this case, and for increasing levels of hydrogen in the products, weight and energy recoveries can be calculated in a similar manner. These results are shown graphically in Figures 9 and 10. The curve for bituminous coal from Figures 7 and 8 are repeated for reference. Point A, calculated from the reported data, lies slightly above the reference curves, since the products in this case still contained significant quantities of nitrogen and sulfur (0.60 and 2.10%, respectively). Point B for the hetero atom-free hydrocarbon product meeting high valued fuel specifications falls below the reference curve. In the real process sufficient char or hydrogen-free carbon could not be generated to produce all of the hydrogen necessary, and additional coal had to be fed to the gasifier. The hydrogen in this coal is not recovered, and the coal's heating value is considerably lower than that of pure coke. Thus the remainder of the curves for these cases falls below the reference curves.

Some design estimates for commercial plants for hydrogenation of Illinois No. 6 coal have been published (3), (11) and from these data weight and energy recoveries can be calculated. Point C on Figures 9 and 10 is based on a plant estimated by Hellwig, et al, while Point D has been calculated from the design developed by O'Hara, et al.

Both "real" cases show much lower efficiencies than the hypothetical curves calculated. Coal as delivered contains both water and ash which must be removed and disposed of. Sulfur and nitrogen must be removed from products in a way which does not give rise to pollution. Hydrogenation units require large amounts of power for compression, pumps, etc., and the generation of this power consumes energy. It is interesting, though perhaps coincidental, that both real cases show essentially the same ratio of their thermal efficiency to that of the reference curves at the same product hydrogen content. The dashed curve through these two points is drawn on the assumption that this same ratio would hold at other levels of hydrogen input.

If new and better technology is developed, thermal efficiencies superior to those shown on the dashed curve in Figure 10 may be realized - although it must be emphasized that the two "real" cases are based on some assumptions which may prove to be optimistic. In any case, the general shape of the curve will not change, and the thermal efficiency of fuel synthesis processes will decrease as fuel hydrogen content increases.

Solid fuels are not well suited to small self-propelled vehicles, and the transportation of fuel from source to point of consumption requires energy. Published studies (12) have indicated that the cost of transporting electrical power is far higher than that of moving an equivalent amount of energy in the form of oil and gas. Thus, it would appear that the conversion of coal to a fluid fuel suitable for pipeline transport should minimize overall energy losses. The fluid fuel

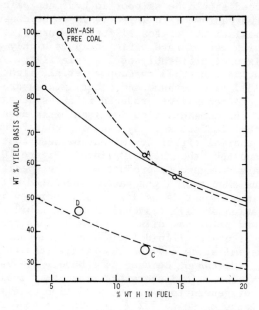

Figure 9.   Weight yields bituminous coal hydrogenation

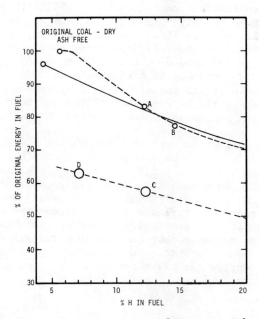

Figure 10.   Energy recovery bituminous coal hydrogenation

which is produced should be as poor in hydrogen as can be
tolerated by the equipment in which it is to be consumed.

To make maximum use of fossil fuel resources, engines should
be designed which can operate efficiently on highly aromatic
hydrocarbon mixture, preferably of low volatility. The stratified charge engine described by Coppoc, et al, might be a
possibility (13). The methane and other light hydrocarbon
gases which must probably be produced in any synthesis plant
should be used for home distribution (synthetic natural gas)
and petrochemical feedstocks. By assuming that future plants
may achieve a thermal efficiency midway between that of the
"real" cases and the "theoretical" line in Figure 10, that
these plants are designed to produce only the specified fuel
and the minimum amount of light gases associated with it, and
that this minimum amount is as low (basis hydrogen input) for
coal hydrogenation at high conversions as is achieved in hydro-
cracking low end point gas oils, the energy loss for manufacturing various types of liquid fuels from coal can be calculated.

Since SO_2 emissions must be restricted, the total heating
value of this coal cannot be used as a base. However, high
sulfur residual fuels (and therefore presumably coal) can be
burned for power generation without pollution problems with a
thermal efficiency of 90 to 95% (14). Thus a base recovery of
93% has been used. A hypothetical plant, producing conventional
gasolines and diesel fuels – 14 %w hydrogen, would lose 28%
of the original energy – having a thermal efficiency of 67%.
The plant would also produce 9 %w light gases. If the liquid
fuel to be manufactured contained only 9.5% H (the level of
xylenes or of an aromatic gas oil), the energy loss would be
reduced to 21%, and the light gas make to 4 %w. To produce
a given amount of liquid fuel, the gasoline-diesel plant would
have to process 15% more coal than the one producing the aromatic
fuel, and would obviously be more expensive. Thus it would
appear that when liquid fuels must be synthesized from coal,
energy resources will be conserved to the greatest extent if
these fuels can be low in hydrogen content – and thus highly
aromatic.

It must be kept in mind that the nature of the raw material
from which fuels are to be synthesized governs the nature of
the fuel which will best utilize the energy available in that
raw material. The manufacture of a highly aromatic liquid fuel
from a hydrogen-rich paraffinic crude oil would require so much
additional processing that the energy recovery would undoubtedly
be lower than if more conventional products were made. But
since it seems probable that future raw materials will be low
in usable hydrogen, aromatic fuel manufacture may represent
one way to conserve natural resources.

Summary

It seems probable that for a considerable time fluid fuels will be desirable at least for self-propelled vehicles, and that these fuels will have to be synthesized from raw materials low in hydrogen, e.g., shale oil, coal or coke. Under these circumstances, fuel composition has a significant influence on the theoretical and practical thermal efficiency of the conversion process, and this affects the fraction of the total energy resources which is available. The manufacture of highly aromatic, low hydrogen content fuels rather than paraffinic materials conserves available energy.

Literature Cited

1. Hinds, G. P., Jr., "Potentials and Constraints on Non-Hydrogenative Processing of Petroleum", Proceedings of the Eighth World Petroleum Congress, Volume 4, pp. 235-244.

2. "Properties of Hydrocarbons of High Molecular Weight Synthesized by Research Project 42 of the American Petroleum Institute 1940-1966", Pennsylvania State University, University Park, Pa.

3. Hellwig, K. C., Chervnak, M. C., Johanson, E. S., Stotler, H. H., and Wolk, R. H., "Convert Coal to Liquid Fuels with H-Coal", Chem. Eng. Progress Symposium Series, 85, Volume 64, (1968), pp. 98-103.

4. Thorne, H. M., Stanfield, K. E., Dinneen, G. U., and Murphy, W. I. R., "Oil Shale Technology - A Review", U. S. Bureau of Mines Information Circular 8216.

5. Cody, W. E. and Seelig, H. S., "Composition of Shale Oil", Ind. Eng. Chem., 44, No. 11 (1952), pp. 2636-2641.

6. Duckworth, J. B., Kane, E. W., Stein, T. W., and Wagner, T. O., "Economic Aspects of Raising Compression Ratio and Octane Quality", Proceedings of the Fifth World Petroleum Congress, Section VI, pp. 91-99.

7. Kavanagh, F. W., MacGregor, J. R., Pohl, R. L., and Lawler, M. B., "The Economics of High-Octane Gasolines", SAE Transactions, Volume 67, (1959), pp. 343-350.

8. Corner, E. S. and Cunningham, A. R., "Value of High Octane Number Unleaded Gasoline in the U. S.", presented before the Division of Water, Air and Waste Chemistry, American Chemical Society, Los Angeles, California, March 28 - April 2, 1971.

9.   Wagner, T. O. and Russum, L. W., "Optimum Octane Number
     for Unleaded Gasoline", presented before the Automobile
     Engineering Meeting of the Society of Automotive Engi-
     neers, Detroit, Michigan, May 14-18, 1973, SAE Preprint
     730552.

10. Ellis, J. C., "Gasolines for Low Emission Vehicles",
     presented before 1973 SAE Section Meetings, SAE Preprint
     730616.

11. O'Hara, J. B., Jentz, N. E., Rippee, S. N., and Mills,
     E. A., "Preliminary Design of a Plant to Produce Clean
     Boiler Fuels from Coal", presented at the American
     Institute of Chemical Engineers 66th Annual Meeting,
     Philadelphia, Pa., Nov. 15, 1973, Paper No. 56E.

12. Linden, H. R., "The Case for Synthetic Pipeline Gas
     from Coal and Shale", Chem. Eng. Progress Symposium
     Series, 85, Volume 64, (1968), pp. 57-72.

13. Coppoc, W. J., Mitchell, B., and Alperstein, M., "A
     Stratified Charge Engine Meets 1976 U. S. Standards",
     presented before the 38th Midyear Meeting of the
     American Petroleum Institute, Philadelphia, Pa.,
     May 16, 1973, Preprint 46-73.

14. Kuhre, C. J. and Sykes, J. A., Jr., "Clean Fuels from
     Low-Priced Crudes and Residues", presented before the
     74th National Meeting of the American Institute of
     Chemical Engineers, New Orleans, La., March 11-15, 1973.

Impact of Automotive Trends and Emissions Regulation on Gasoline Demand

DAYTON H. CLEWELL

Mobil Oil Corp., New York, N. Y. 10017

WILLIAM J. KOEHL

Mobil Research and Development Corp., Paulsboro, N. J. 08066

Gasoline in Perspective

In these days of energy shortages, gasoline is probably our most visible source of energy and the automobile, our most conspicuous consumer. In this paper we will trace some automotive trends that have contributed to rising gasoline demand in the past and that can be expected to affect it in the future. We will also point out some opportunities for moderating the future growth of gasoline demand.

To put gasoline consumption into the perspective of the overall energy picture, Figure 1 shows total 1971 energy consumption broken down into its major sources and consuming sectors(1). On an equivalent energy basis, gasoline represented about 17% of the energy consumed in 1971, or about 38% of the oil. Virtually all of the gasoline was consumed by the transportation sector, but not all by automobiles. Gasoline supplied about 68% of the energy required for all transportation.

Focusing on the automobile, statistics illustrated in Figure 2 show that gasoline consumption by passenger cars amounted to 74% of total gasoline demand(1,2) or about 13% of total energy.

Light trucks, up to 6000 lbs. gross vehicle weight, are estimated to consume about 10% as much gasoline as automobiles, or 7% of the total(3). Although they frequently are used interchangeably with automobiles, we will exclude them from this discussion because they are treated separately from

---

*Numbers in parentheses designate references at end of paper.

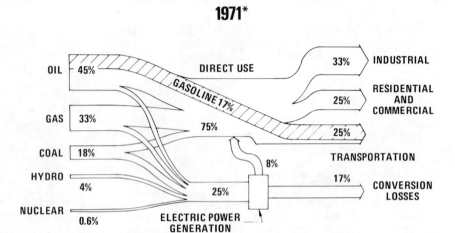

U.S. Department of Transportation

Figure 1.   Patterns of fuel and energy consumption

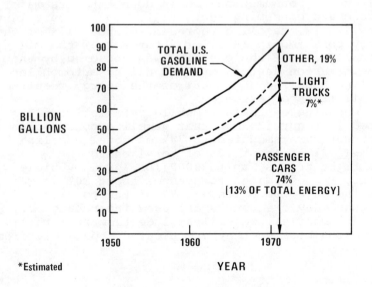

*Estimated

Figure 2.   Gasoline consumption, 1950–1971

passenger cars in emissions regulations for 1975 and beyond, and because they appear to be excluded from proposals under discussion in Washington for regulating fuel economy.

Car Trends

The increase in gasoline demand shown in Figure 2 reflects changes in passenger car sales and usage. Based on data from the U.S. Department of Transportation(4), Figure 3 shows that registrations increased 45% from 1960 to 1970 while miles traveled were up slightly more at 51%. Gasoline consumption by passenger cars increased 60% indicating an increase in average fuel consumption per mile over the period.

Over the same 1960 to 1970 period (actually, data are shown through 1972) there were changes in several aspects of new car design that caused fuel consumption to increase. As illustrated in Figure 4, vehicle weight(5), engine displacement(6), and the percentage of automatic transmissions(7) all increased, but not by large factors. At the same time these fuel consuming trends were offset somewhat by the fuel saving trends of increasing compression ratio(6) and greater sales of imported cars(8). Only air conditioner installations(7) showed a steep increase, but the resulting increase in fuel consumption was at a much lower rate because air conditioning is not used all of the time.

In addition to the trends shown in Figure 4, exhaust emission controls emerged with the 1968 models (1966 for cars sold in California) as a new aspect of car design. These have had an increasingly important effect on gasoline demand; by 1970 they had reduced fuel economy of new cars by about 3%. For the 1973 models, we estimate that the total loss in fuel economy caused by emission controls averaged about 15%.

In the future, car design trends will continue to influence average fuel economy and total gasoline consumption. We expect emission controls to continue to be an important factor.

Of the factors traced in Figure 4, weight could increase because of safety and damageability regulations; however, this trend may be counteracted by increasing emphasis on weight reduction wherever possible to improve fuel economy. On the average, we expect vehicle weight to decrease because of the growing proportion of smaller cars among new car sales. Automatic transmission installations appear

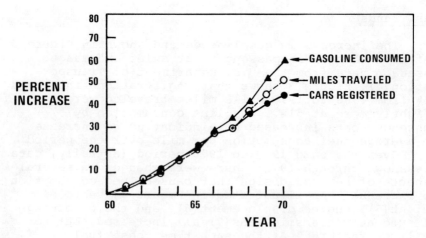

Figure 3. Passenger car trends, 1960–1970

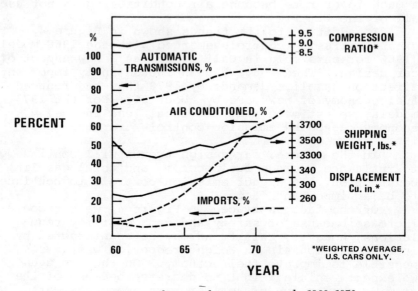

Figure 4. New car design and equipment trends, 1960–1972

to have leveled off and air conditioner installations probably cannot increase much more without deep penetration into the small car and cooler climate markets.

Compression ratio has already suffered a drastic reduction since 1970. This change, made to facilitate introduction of unleaded gasoline for 1975 catalytic emission control systems, accounts for nearly half of the 15% fuel economy loss associated with emission controls through 1973. Some further reduction in compression ratio could occur in the next year or two to provide adequate knock protection in cars using 91 octane unleaded gasoline. In the longer term, after emission control development has stabilized, compression ratio could be increased to regain some economy and performance; however, this would require premium grade higher octane fuels. Engine displacement will probably increase somewhat in given car families to compensate for performance losses due to emission controls, but on average, it may decrease because of the shift to smaller cars.

Later in this paper we shall make some projections of growth in future gasoline demand as a function of the two variables that we see as most important in the short term - emission control regulations and average vehicle weight. The latter will be described in terms of the shift from larger cars to smaller ones that can be seen developing rapidly in the new car market today. Projections of the potential effects of emission regulations are also timely, not only in the context of this symposium, but also in the context of current deliberations on possible changes in the Clean Air Act.

Gasoline Trends

Before projecting gasoline demand, let us look briefly at gasoline quality trends. These also have been and will continue to be influenced by vehicle trends. Along with compression ratio, gasoline octane quality reached its highest levels in the latter half of the last decade; while gasoline volatility showed relatively little change (9).

Since 1970, however, important changes in gasoline quality have begun to take place in response to emissions regulations. Several marketers introduced unleaded grades of lower octane than traditional regular in 1970 and 1971 as new cars with lower compression ratio engines were introduced. By July 1, 1974, virtually all marketers must offer an unleaded gasoline of at least 91 research octane number (RON)

in anticipation of catalyst equipped 1975 cars (10).
As cars requiring unleaded gasoline replace today's
cars which can use leaded gasoline, the average octane
quality of the gasoline pool without lead will have to
increase from recent levels of approximately 88 RON to
91 RON.

Recent regulations will accelerate the increase
in the clear pool octane level by forcing a reduction
in the lead content of existing regular and premium
grades. EPA has ordered that the lead content of the
leaded grades of gasoline be reduced from their recent
average levels of about 2 to 2.5 g/gal to a level such
that the average lead content of all grades including
the unleaded gasoline will not exceed 1.7 g/gal after
January 1, 1975. The limit will be reduced in four
more steps to 0.5 g/gal by January 1, 1979 (11).

Turning to volatility, we do not expect any large
changes in the future. Although some had been proposed
to minimize cold start emissions from future low
emission cars, we have not seen convincing evidence
that this is the best alternative for effective
emission control. Advances are being made in
carburetor and manifold design to facilitate fuel
evaporation and induction, and accomplish the same
objective. Today's gasoline volatility evolved
through many years of optimization of vehicle performance and efficient utilization of gasoline
blending stocks. Any drastic change in the distil-
lation specifications, particularly reducing the
higher boiling components of gasoline, would require
wasteful changes in refining schemes. It would deny
refiners the flexibility to make optimum use of
available blending components and would require
conversion of less volatile stocks to more volatile
ones with attendant yield losses and increases in
crude runs and costs. This would be unadvised in the
present energy limited environment.

Projecting Future Gasoline Demand

In order to relate the car trends discussed in
this paper to gasoline demand, we have devised a
mathematical model in which the car population is
described by means of parameters which are related
to fuel economy. The car population is treated as
being composed of five classes of cars: full size,
intermediate, compact, domestic sub-compact, and
imported. Historical data are used to determine the
fraction of total new car sales in each class in each
model year, and also to calculate sales weighted

averages for the parameters used to describe each class. Projections of future sales fractions and of values of the parameters are made from the historic bases.

The main determinant of demand growth in the model is the growth in annual vehicle miles. Changes in the car parameters and sales shifts between car classes cause perturbations on the basic demand growth curve. For the growth curve of annual vehicle miles, we have used the projection of the Department of Transportation which was given in the January 1972 report of the Committee on Motor Vehicle Emissions (12). This curve, illustrated in Figure 5, may now be less valid because of fuel shortages and price increases. Still it does not seem unreasonable when it is compared with census projections. It projects growth in vehicle miles at an average of 3.4% per year through 1980 and 1.9% from 1980 to 1990; while, the labor force is projected to increase at 1.6% and the number of households at 2.2% per year from 1970 to 1990 (13). In any case, it will serve as a basis on which to assess the relative impacts of potential changes within and among car classes.

To calculate total fuel consumption in any year, the average fuel economy of each model year in the car population and the miles traveled by cars of each model year must first be obtained. Total annual miles are obtained from Figure 5, described above. The fraction of the miles traveled by each model year in the population is taken from a Department of Transportation estimate of relative annual miles versus car age (12). The average fuel economy of each model year is calculated using the five class breakdown of car sales in that model year and the parameters describing each class.

This mathematical model was used to project gasoline demand through 1985 and to explore potential effects of emissions limits, car size, and possible efficiency improvements on the growth in future gasoline demand.

Potential Effects of Emission Controls

Based on published information, primarily statements by the automobile manufacturers at recent Congressional hearings (see Appendix) and on our own engineering assessment of the control systems likely to be used in the coming years, we believe that emission control regulations will lead to the fuel economy changes shown in Figure 6. Compared with uncontrolled

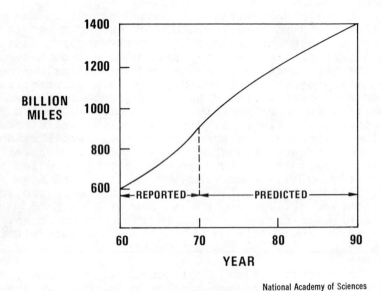

National Academy of Sciences

Figure 5.   Growth in vehicle miles traveled/year

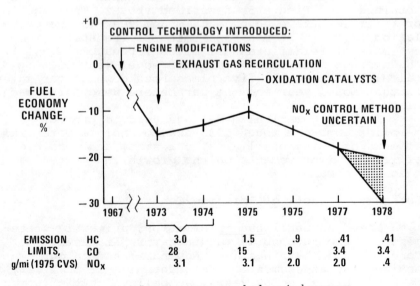

Figure 6.   Effect of emission standards on fuel economy

cars of 1967, we see a 15% loss in fuel economy in the 1973 models. Some improvement appears to have occurred in 1974, and in 1975 models outside of California, we expect a 3% improvement over 1974.

In Figure 6, the dates assigned to the standards shown beyond 1975 are the earliest model years in which we would expect them to apply in view of recent actions in Congress (22).

The loss through 1973 is attributed to changes in spark timing and air-fuel ratio, and to introduction of exhaust gas recirculation, changes made to control pollutant formation in the engine, as well as to reduction of compression ratio in anticipation of 91 octane unleaded gasoline. With the introduction of catalysts in the exhaust stream to control hydrocarbon and carbon monoxide emissions from cars meeting the 1975 49-states interim standards, engine calibrations can be partially restored to more efficient settings to regain a small part of the fuel economy lost because of emission controls. However, beyond these 1975 limits, fuel economy losses will again increase and could rise to 30% if the most stringent standards set by the Clean Air Act of 1970 must be met. Our projection of diminishing fuel economy at the more stringent emission levels beyond 1975 parallels published estimates by General Motors Corporation and Ford Motor Company (see Appendix).

With the 1975 interim California standards and the standards indicated for 1977, we expect the fuel economy losses relative to uncontrolled cars to be 14 and 18%, respectively. These are associated with retarded spark timing, richer air-fuel ratios, and exhaust gas recirculation required primarily for NO_x control.

Least certain of attainment are the statutory limits of 0.4 g/mi. for hydrocarbons, 3.4 for carbon monoxide, and 0.4 for oxides of nitrogen ultimately required by the Clean Air Act. Because adequate technology has not yet been identified for meeting these standards, our estimates of the fuel economy penalty ranges from 20% to as much as 30% relative to 1967 cars. Our optimistic projection assumes development of a nitric oxide reduction catalyst which cannot presently be regarded as technically feasible in the context of the regulations; while our pessimistic projection assumes NO_x control could be achieved through the maximum degree of exhaust gas recirculation that is consistent with engine operation. Some auto industry projections for the nitric oxide reduction catalyst are closer to our pessimistic estimate.

In this assessment, we have tried to isolate the effect of engine changes made primarily for emission control. Compression ratio reduction is included among these changes. We have also endeavored to compare systems on a constant performance basis. In evaluating vehicle test data, one must be careful to separate emission control effects from counteracting effects of other changes, such as reduced acceleration performance, high energy ignition systems, improved carburetors, radial tires, or shifts to smaller cars. These changes would yield fuel economy improvements regardless of the emission control system chosen.

To compare the potential effects of the emission limits shown in Figure 6, we have used the corresponding fuel economy changes in our math model. The new car sales mix (full size, intermediate, compact, sub-compact, and imported) in future years was assumed to be the same as in 1973, while the parameters describing the new cars in each future year were also held at current values. Annual vehicle miles were allowed to increase as projected in Figure 5. The projected growth in gasoline demand through 1985 at each emission level is traced in Figure 7. For all of these projections, the gasoline is assumed to be 91 octane unleaded gasoline.

The fuel economy debits attributed to each emission level were held constant through the period of the projections in Figure 7. Evolutionary improvements can reasonably be expected that would reduce the fuel economy debits as manufacturing and service experience grows; however, such improvements ought to be applicable in some degree at all of the emission levels so that on a relative basis, our gasoline demand projection would not change significantly.

The gasoline demands projected in Figure 7 for the year 1985 are summarized in the table below along with the percentage changes in 1985 from a base case in which 1973/74 emission standards are assumed to apply to all model years through 1985. In computing the demand for any year, each emission limit is applied from its year of introduction until it is superseded by a subsequent, more restrictive standard.

Emission Limits		1985 Projection of 91 Octane Unleaded Gasoline Required for Automobiles	
Year of Introduction	HC, CO, NO$_x$*	billion gal.	Change from Base, %
1973	3, 28, 3.1	121	Base
1975 49-States	1.5, 15, 3.1	117	− 3
1975 California	.9, 9, 2	122	+ 1
1977 Proposed	.41, 3.4, 2	128	+ 6
1978 Statutory	.41, 3.4, .4	132 to 150	+ 9 to +24

*g/mi. on 1975 CVS test basis.

For perspective, the maximum demand increment with the most severe emissions limits (29 billion gallons annually or 1.9 million barrels per day) is greater than the gasoline shortfall of 1.4 million barrels per day anticipated during the embargo on shipment of Arabian oil to the United States in the Winter of 1973-1974 (3).

While the fuel saving projected for the 1975 system relative to present controls may be less than projected by others, the important point is that moving to more stringent emission limits than the 1975 interim limits will require substantially more gasoline. Therefore, it seems reasonable to set emission standards at levels that are justifiable in terms of attaining desired ambient air quality levels, not at arbitrary levels that are unduly restrictive and demanding of energy.

Prospects for Moderating Gasoline Demand Growth

Even with the lowest demand growth curves in Figure 7, which are based on new car populations as in 1973, we project an increase in gasoline demand for automobiles of almost 50% from 1973 to 1985. This can be compared with the 50% growth from 1970 to 1985 projected by the Transportation Energy Panel (15) and with growth in total energy requirements projected by others. For example, the National Petroleum Council projected growth of 70 to 90% from 1970 to 1985 (14) and the Bureau of Mines projected about 70% over a similar period (16).

What are the prospects for substantially reducing, or even avoiding, this projected 50% growth in gasoline demand for automobiles? With the technology available now, two moderating trends are obvious possibilities, and both could be seen to be taking some effect in recent months. One is reduction in vehicle miles traveled, the other is replacement of large cars with smaller ones. In the longer term, improvements in automotive technology can be expected to provide additional fuel economy benefits. These will come through improved engine and drive train efficiencies, lighter weight components, improved efficiency in power accessories, and improvements in body aerodynamics. We will explore a few of these options.

Reduced Vehicle Miles. In recent months we have seen a forced reduction in car usage because there has not been enough gasoline available to let us drive as much as we would like. People are adjusting to life with less gasoline and less car travel, but not without inconvenience and even hardship. Since even recreational driving is important to the person accustomed to it, and especially to the person whose livelihood depends on it, car usage can be expected to return to more normal levels with increasing supplies of gasoline now becoming available. In the longer term gasoline prices will influence people to switch to more efficient cars that will enable them to drive the required miles. Consequently, there is unlikely to be a reduction in annual car miles which, by itself, would have a substantial effect in reducing gasoline demand.

Opportunities for some mileage reductions appear to exist in increased car pooling and the use of public transportation. The Department of Transportation estimates that potential savings from increased car pooling, and increased usage of urban transit, and intercity bus and rail would be only 1.2, 1.0 and 0.5%, respectively, of national transportation fuel requirements after five years (3). These savings could increase to 8.3, 1.7, and 1.3%, respectively, in 15 years (3).

Shift to Smaller Cars. Using our math model, we have estimated the potential 1985 gasoline demand with a shift of new car sales from large cars to smaller ones as compared to maintaining the 1973 new car sales mix. The base 1973 new car sales mix and a future new car mix achieved with increasing penetration of small cars are shown in Figure 8 in terms of the car classes

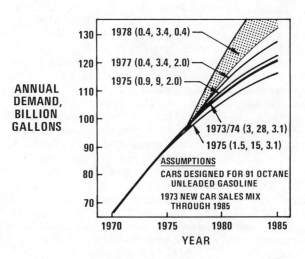

*Figure 7.   Gasoline demand projections for automobiles
—effect of emission standards*

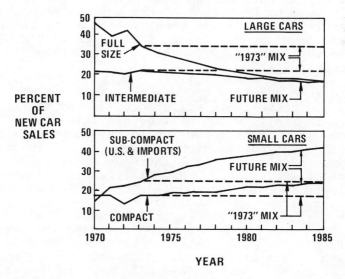

*Figure 8.   Hypothetical future new car sales mixes for projecting
gasoline demand*

used in our model. Figure 9 shows the gasoline demand
growth curves calculated for the two cases. It can be
seen that if the postulated shift to smaller cars were
achieved, gasoline consumption by automobiles would
increase only about 37% over that calculated for 1973
versus 51% if the 1973 new car sales mix were maintained. This represents a potential saving in 1985 of
9%. For this comparison, 1973/74 emission controls
were assumed to be in effect through 1985. With 1975
emission controls, the saving would be about the same,
but the gasoline demand of the postulated future mix
would be about 3% lower than shown in Figure 9.

The full benefit of the postulated shift to
smaller cars would not be seen until 1990 or later
when nearly all pre-1980 cars would be replaced.
Figure 9 also shows the projected fuel savings associated with the shift in sales mix that occurred from
1970 to 1973 as the Pinto, the Vega, and the Gremlin
were introduced. With the 1970 new car mix adjusted
to the same emission levels as the 1973 mix, this
would represent an estimated saving of 6% in 1985.

Our postulated reduction of the big car segment
(full size and intermediate) of the new car market
from about 58% in 1973 to 40% by 1980 does not seem
unreasonable in light of the reduction from 68% in
1970 that has already occurred. According to recent
news stories, Ford has indicated that small cars
could amount to 50 to 60% of the market as soon as
September, 1974(17); and General Motors has indicated
that small car production capability in 1975 would be
30% greater than in 1974 and could be doubled in 1976,
if necessary(18).

Improved Vehicle Efficiency. Improvements in
engine and drive train efficiency (excluding increased
compression ratio which will be discussed later)
coupled with weight reduction in car components and
improved aerodynamics might reasonably be expected to
afford a further 10 to 15% increase in fuel economy
when fully implemented. Radial tires and improved
efficiency in power accessories could also contribute
to this saving. Assuming that these improvements lead
to 10% better fuel economy in 1980 models and 15% in
1985 models, we project a potential gasoline saving
in 1985 of 11%. This estimate is based on 1973/74
emission levels and the shift to smaller cars postulated above. As in the case of the shift to smaller
cars, the full 15% benefit of the postulated improvements would not be realized until all less efficient
cars are replaced.

Our postulated 10 to 15% efficiency improvement is also corroborated by Ford and General Motors predictions that the minimum fuel economy of future standard size cars is likely to be 13 to 15 mpg (18); as contrasted to 11.5 mpg and less for the average 1970 to 1974 full size cars in our model.

As stated earlier, we estimate that about half of the fuel economy loss associated with emission controls (7 of the 15% for 1973 systems) is due to the reduction in compression ratio that was made to accommodate 91 octane unleaded gasoline. This energy loss could be recovered in part by use of higher octane unleaded gasoline in engines of increased compression ratio. It could be recovered in full, and even beyond, if the current emission standards were retained, or if lead tolerant low emission engines or emission control systems were developed to meet more stringent standards. Then leaded gasoline could continue to be used, and compression ratio could be increased.

Of course, the recently published regulations (11) requiring a phase-down of the lead content of gasoline would have to be revoked to allow the full fuel saving potential of lead to be realized. If air quality standards for lead particulates were then established, devices to remove lead from vehicle exhaust could be applied to meet them. Several companies have reported very marked progress in the development of these devices which shows that they are feasible for 80% removal of lead particulates from exhaust (19,20).

Our studies indicate that adding 2.5 g of lead per gallon to the 91 octane unleaded gasoline that will soon be available because of current regulations (10, 11) would permit an octane number increase to 98 and an engine efficiency increase of up to 13%.

At the 1973/74 emissions level with the 1973 new car sales mix, the 1985 gasoline demand of 121 billion gallons could be reduced by up to 16 billion gallons, if the full fuel saving potential of leaded gasoline were exploited by that time.

In our postulated scenario of increasing penetration of smaller cars along with other fuel economy improvements, the added effect of lead in moderating the growth in gasoline demand could be as shown in Figure 10. For this projection, 1973/74 emission standards were assumed to apply along with the 10 to 15% efficiency improvement discussed above. The compression ratio increase which would be equivalent to a 13% increase in efficiency was assumed to be phased in from 1976 through 1979. The resultant saving

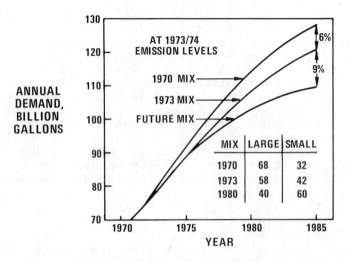

Figure 9. Gasoline demand projection for automobiles—effect of
new car sales mix

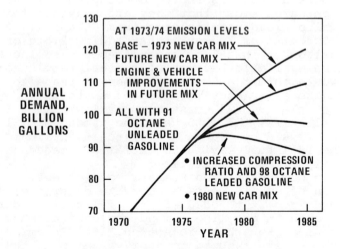

Figure 10. Gasoline demand projections for automobiles—
effect of efficiency improvement

in 1985 could be 10% but the full benefit would not
be realized until sometime later when all lower
compression ratio cars could be replaced.

Summary

Based on past trends, gasoline demand has been
projected to increase about 50% from 1973 to 1985 at
current or 1975 49-states emission levels, and maybe
as much as 90% if the statutory NO_x emission standard
of 0.4 g/mi. is retained and can be approached only
by EGR. These projections assume that the mix of
vehicle types on the road will remain the same as in
1973 and that future car usage will grow as projected
before the current gasoline shortage. However, this
rate of increase in demand is not inevitable. Forces
are already at work that could moderate it. The
presently evident trend toward smaller cars, if it
continues, along with development of technology to
implement the improvements in vehicle efficiency that
appear feasible could substantially reduce the rate
of growth in gasoline demand. Choosing vehicle emission standards with due concern for attainment of air
quality standards and conservation of fuel should
prevent unnecessary increases in demand.
Potential 1985 gasoline savings projected in this
paper are summarized below. If it were possible to
implement all of them, automobiles meeting the 1973/74
emission standards and using leaded gasoline might
require only 10% more gasoline in 1985 than calculated
for 1973; while those meeting 1975 interim 49-states
standards with unleaded gasoline might require only
19% more. At more stringent standards, the growth
would be greater.

Emission Level HC, CO, NO_x, g/mi.	1973/74 3,28,3.1	1975 49-States 1.5,15,3.1
1985 Demand with Uncon- strained Growth, bil. gal. Increase over 1973 Demand of 80 bil. gal.	121 51%	117 46%
1985 Saving with Shift to Smaller Cars, bil. gal.	11.3	10.8
1985 Saving with Improved Engine and Drive Train Efficiency and Vehicle Weight Reduction, bil. gal.	12.0	11.3
1985 Saving with Leaded Gasoline and Increased Compression Ratio, bil. gal.	9.5	Not available if catalysts are used
1985 Demand if all Savings are Realized, bil. gal. Increase over Calculated 1973 Demand of 80 bil. gal.	88.2 10%	94.9 19%

Our optimistic projection that the 1985 gasoline demand for automobiles alone could be held to 10 to 20% more than in 1973 is dependent, as discussed above, on changes in promulgated emissions regulations, continuation of the trend to small cars, and other improvements in fuel economy. It would represent a substantial potential gasoline saving and would be equivalent in energy to about two million barrels per day less crude oil than projected by the uncon- strained demand curve. Compared to total projected 1985 petroleum requirements of 25 million barrels per day (14,21), it represents a saving of about 8% and relative to total 1985 energy projections of 58 million barrels per day oil equivalent (21), a saving of about 3%. These last comparisons show that while the projected gasoline saving could be very substantial in absolute terms, it would still be a relatively small saving in total energy requirements. Therefore, while highly desirable, it should not be viewed as the complete answer to our national energy problems. Con- servation in other consuming sectors is also needed.

In adopting energy conservation policies for the future, we should remember that liquid hydrocarbon fuels are ideally suited for transportation applications. Current shortages are due not only to rising demand for all fuels but also to substitution of petroleum fuels for higher sulfur coal and for scarce gas. To conserve petroleum and minimize the economic and political costs of importing oil, we should try not only to moderate the rising demand for fuels but also to reserve liquid petroleum fuels for premium applications such as transportation while finding environmentally acceptable ways to use domestic coal and nuclear energy in applications such as power generation that are adaptable to a variety of fuels.

Acknowledgment

We gratefully acknowledge the able assistance of Drs. C. R. Morgan and M. Ohta of Mobil Research and Development Corporation who developed the mathematical model used in this study.

Literature Cited

1. "Energy Statistics, A Supplement to the Summary of National Transportation Statistics," U. S. Department of Transportation, Report No. DOT-TSC-OST-73-34, September, 1973.

2. Annual Statistical Reviews, U. S. Petroleum Industry Statistics, American Petroleum Institute; and, 1972 National Petroleum News Factbook Issue, and earlier issues.

3. Whitford, R. K. and Hassler, F. L., "Some Transportation Energy Options and Trade-Offs: A Federal View," presented at California Institute of Technology, January 8, 1974.

4. Highway Statistics - 1960 through 1970 issues, U. S. Department of Transportation, Federal Highway Administration, Bureau of Public Roads.

5. Calculated using shipping weight data from Automotive Industries Statistical Issue, 1960 through 1972, and Mobil estimate of car population distribution.

6. Brief Passenger Car Data, Ethyl Corp., 1960 through 1970; 1971 and 1972 National Petroleum News Factbook Issue.

7. Ward's Automotive Year Book, 1960 through 1970;
    Ward's Automotive Reports, September 25, 1972.

8. Ward's Automotive Reports, February 26, 1973.

9. Mineral Industries Surveys, Motor Gasolines,
    U. S. Department of the Interior, Bureau of Mines.

10. Regulation of Fuels and Fuel Additives, Federal
    Register, January 10, 1973, p. 1254.

11. Regulation of Fuels and Fuel Additives, Federal
    Register, December 6, 1973, p. 33734.

12. Semiannual Report by the Committee on Motor
    Vehicle Emissions, National Academy of Sciences
    to the Environmental Protection Agency,
    January 1, 1972.

13. A Guide to Consumer Markets 1973/74, Report No.
    607, The Conference Board, Inc., New York, N. Y.,
    1973.

14. U. S. Energy Outlook, A Summary Report of the
    National Petroleum Council, December 1972.

15. "Research and Development Opportunities for
    Improved Transportation Energy Usage - Summary
    Technical Report of the Transportation Energy
    R&D Goals Panel," U. S. Department of Transpor-
    tation, September 1972.

16. The Potential for Energy Conservation: A Staff
    Study, Executive Office of the President, Office
    of Emergency Preparedness, October 1972.

17. Automotive News, January 21, 1974, p. 1.

18. Automotive Industries, March 1, 1974, p. 17 and 18.

19. Cantwell, E. N., Jacobs, E. I., Kunz, W. G. and
    Liberi, V. E., Society of Automotive Engineers,
    National Automobile Engineering Meeting, Detroit,
    Michigan, May 1972, Paper 720672.

20. Hirschler, D. A., Adams, W. E. and Marsee, F. J.,
    National Petroleum Refiners Association Annual
    Meeting, San Antonio, Texas, April 1-3, 1973,
    Paper AM-73-15.

21. DuPree, W. G., "United States Energy Requirements to the Year 2000," American Society of Mechanical Engineers, presented at Intersociety Conference on Transportation, Denver, Colorado, September 1973, Paper No. 73-ICT-105.

22. S2589 (Energy Emergency Act) Conference Report, December 1973.

APPENDIX

Effect of Emission Controls on Fuel Economy

In Figure A-1, our assessment of the fuel economy losses associated with emission controls is compared with published estimates from several sources.

The Ford report of a 13% loss in city-suburban fuel economy in 1973 relative to uncontrolled cars is based on analysis of a large number of cars in a fleet whose composition was constant through the period (A-1). Effects of emission controls were separated from effects of changes in vehicle weight, engine displacement, rear axle ratio, and accessory equipment. Projections for the future years are made from this 1973 base (A-2). Note that Ford expects a 3% gain from 1974 to 1975 interim 49-states standards in catalyst-equipped vehicles but a 5% loss in vehicles without catalysts.

The Chrysler estimate for 1973 is based on an analysis of a typical intermediate size car (A-3), while those for 1974 and 1975 are based on testimony to the Senate Subcommittee on Public Works (A-4).

General Motors' statement to the Senate Committee on Public Works indicates a 15% loss due to emission controls in 1973 relative to uncontrolled cars, after weight effects are subtracted (A-5). From 1973 to 1974, it shows a 1% gain based on the average of typical high volume models [the point (x) in Figure A-1], but a 1% loss based on a sales weighted average after correcting for vehicle weight increases. In going from 1974 to 1975 interim 49-states standards, GM estimates range from 10% to about 20%. The GM statement shows a 13% improvement in city traffic economy on an average basis at an equivalent vehicle weight. In oral testimony GM indicated that the improvement attributable to the emission controls alone was only 10% (A-6). In controlled tests with one large

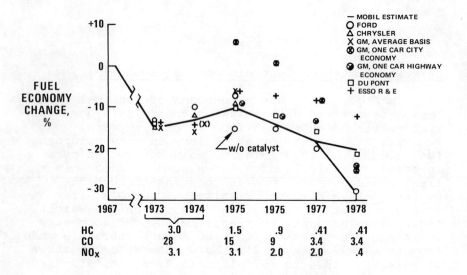

Figure A-1.   *Effect of emission standards on fuel economy*

car, a 21% improvement was indicated in city fuel economy relative to an average for new 1973 production models of this car; for highway economy the improvement was only 6% (A-5). In Figure A-1, we have plotted both the 10% estimate and the one-car results to establish the range of GM estimates for the 1975 49-states standard. For more severe standards than the 1975 49-states standard, the one-car results show the downward trend in fuel economy.

The DuPont assessment (A-7) for 1973 is based on measurements with a constant fleet of six production cars. It is adjusted for weight changes. For both 1975 emission levels the estimates are based on the extent to which carburetion and spark timing could be changed to improve economy while still meeting the NO_x standards. The 1977 and 78 points are based on tests with cars equipped with low thermal inertia manifolds, oxidizing catalysts for 1977 and both oxidizing and reducing catalysts for 1978.

The Esso Research and Engineering Company estimates (A-8) have been used by Russell Train, Administrator of the EPA, in a presentation to the House of Representatives (A-9). These estimates appear to be based on optimistic expectations for the efficiency of reducing catalysts and a theoretical assessment of optimized exhaust recycle systems which involve advancing spark timing, decreasing air-fuel ratio, and limiting EGR rate in proportion to engine air.

The Mobil estimate for 1973 includes fuel economy debits for compression ratio reduction, retarded spark timing and exhaust gas recirculation (A-10). For the 1975 49-states standard, it is somewhat lower than the average of the published estimates shown in Figure A-1, because we expect that compression ratio will be reduced further or spark timing will be retarded to provide an acceptable level of octane satisfaction in 1975 cars using 91 octane unleaded gasoline. These changes would reduce fuel economy. Beyond the 1975 interim 49-states standard, our estimate of decreasing fuel economy parallels most of the published estimates shown in Figure A-1.

REFERENCES FOR APPENDIX

A-1.   LaPointe, C., "Factor Affecting Vehicle Fuel Economy," Society of Automotive Engineers National Meeting, Milwaukee, Wisconsin, Sept. 1973, Paper No. 730791.

A-2.   Statement of H. L. Misch, Ford Motor Co., to
       the U. S. House of Representatives, Subcommittee
       on Public Health and Environment, Dec. 4, 1973,
       Washington, D. C.

A-3.   Huebner, G. J. and Gasser, D. J., "Energy and
       the Automobile - General Factors Affecting
       Vehicle Fuel Consumption," Society of Automotive
       Engineers, 1973 National Automobile Engineering
       Meeting, Detroit, Michigan, May 1973, Paper
       No. 730518.

A-4.   Riccardo, J. J., Chrysler Corp., Testimony before
       the U. S. Senate Committee on Public Works,
       Nov. 5, 1973, Washington, D. C.

A-5.   "Fuel Economy vs. Exhaust Emissions," Attachment
       7 of General Motors Corp. Statement submitted to
       the U. S. Senate Committee on Public Works,
       Nov. 5, 1973, Washington, D. C.

A-6.   Cole, E. N., General Motors Corp., Testimony
       to the U. S. House of Representatives, Subcom-
       mittee on Public Health and Environment,
       Dec. 4, 1973, Washington, D. C.

A-7.   Cantwell, E. N., "The Effect of Automotive
       Exhaust Emission Control Systems on Fuel
       Economy," E. I. duPont de Nemours and Co.,
       Petroleum Laboratory, Jan. 14, 1974.

A-8.   Personal communication - L. E. Furlong, Esso
       Research and Engineering Co.

A-9.   Train, R. E., Statement and Attachments
       presented to the U. S. House of Representatives,
       Subcommittee on Public Health and Environment,
       Dec. 3, 1973, Washington, D. C.

A-10. Clewell, D. H., Mobil Oil Corp., Addendum to
       Testimony before the U. S. Senate Committee on
       Public Works, Nov. 5, 1973.

3

Gaseous Motor Fuels—Current and Future Status

W. R. FINGER —Humble Oil and Refining Co., Baytown, Tex. 77520

D. S. GRAY—American Oil Co., Whiting, Ind. 46394

W. J. KOEHL—Mobil Research and Development Corp., Paulsboro, N. J. 08066

P. E. MIZELLE—Cities Service Oil Co., Cranbury, N. J. 08512

A. V. MRSTIK—Atlantic Richfield Co., Harvey, Ill. 60426

S. S. SOREM—Shell Oil Co., San Francisco, Calif. 94104

J. F. WAGNER—Gulf Research and Development Co., Harmarville, Pa. 15238

Some of the early versions of implementation plans for achieving and maintaining the national ambient air quality standards included provision for the mandatory conversion to gaseous fuels of certain groups of motor vehicles. Also, in the State of California, there were legislative proposals for similar retrofit programs. Because of the widespread interest in this subject, the Engine Fuels Subcommittee of the American Petroleum Institute Committee for Environmental Affairs undertook an assessment of published information on the potential impact of adoption of such a gaseous fuels retrofit program. This assessment has been published and is available in its entirety as an API Publication. (1) This presentation summarizes the API publication.

To demonstrate the incentive for retrofit to gaseous fuels, a typical set of emission curves will be presented. (2) Figure 1 shows the total hydrocarbon emission from an engine operated at one-half throttle on three different fuels: gasoline, propane and natural gas. The reduction in the hydrocarbon emission which can be achieved by switching to a gaseous fuel is the primary incentive for the retrofit proposals. Not only are the hydrocarbon emissions nearly cut in half as we switch from gasoline to the gaseous fuels but the emitted hydrocarbons tend to be less reactive when gaseous fuels are used. (2), (3), (4), (5) Further, note that as we approach the lean misfire limits the hydrocarbon emissions start to increase rapidly. With the gaseous fuels, the air/fuel equivalence ratio at which we get into lean misfire problems is considerably farther out and thus we can operate at

43

leaner mixtures without increases in total hydrocarbon
emission. A reason why we may wish to operate at
leaner mixtures is illustrated in Figure 2 which shows
the nitrogen oxide emissions as a function of air/fuel
equivalence ratio. Here we see that substantial
reductions in nitrogen oxides emissions can be achieved
without any significant penalty with respect to hydrocarbon emissions by utilizing the lean operating
capability of the gaseous fuels.

Figure 3 shows the carbon monoxide emission as a
function of the air/fuel equivalence ratio. Here you
will see that the emission of this pollutant is substantially unaffected in the operating range of
interest. Shown in the same Figure is the horsepower
output as a function of the air/fuel equivalence ratio.
You will note that these half-throttle data indicate
that there will be some sacrifice in power output
operating at the lean equivalence ratios. Because of
the definition of half-throttle used in this investi-
gation, the horsepower data for the separate fuels
are not sufficiently different to justify plotting
separate lines. However, at full-throttle, there is
a definite power loss upon converting to gaseous fuels.
On propane conversions power losses usually less than
10% have been reported. (3), (6), (7) On natural gas
conversions the power losses have generally been in
the 10% to 15% range, (3), (7), (8) all relative to
gasoline. These power losses are due primarily to the
fact that at full throttle the maximum charge of com-
bustible mixture entering the engine cylinder is
essentially a constant number of moles if all of the
fuel is in the gaseous state. With the lower mole
weight fuels the energy per unit volume of charge is
less. (4) Gasoline/air mixtures not only have the
highest average molecular weight but some portion of
the gasoline may actually enter the combustion chamber
in the form of liquid droplets, thus further increasing the maximum quantity of energy producing charge
which may be introduced into the cylinder.

Having established an incentive for a gaseous fuel
retrofit program we will next look at what is involved
in retrofitting a vehicle. Table 1 gives some of the
pertinent properties of the three fuels in question.
For this presentation the simplification has been made

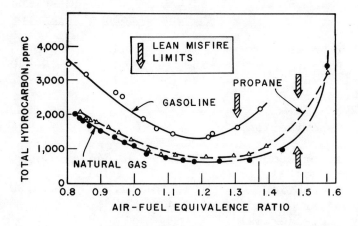

Figure 1.   Hydrocarbon emissions as a function of air–fuel equivalence ratio at 50% throttle

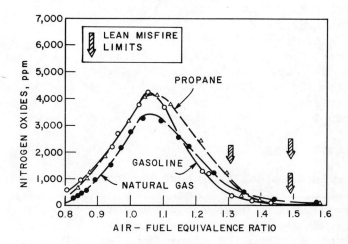

Figure 2.   Nitrogen oxide emissions as a function of air–fuel equivalence ratio at 50% throttle

Table I

Physical Properties of Methane, Propane, and Gasoline

Property	Methane	Propane	Gasoline (approx)
Molecular Weight	16	44	100
Boiling Point, ^{O}F	-259	-44	100-400
Critical Temp, ^{O}F	-116	207	--
Vapor Pressure @ 100°F, psi	--	190	8-14
Heat of Vaporization, BTU/lb	220	184	--
Gross Heat of Combustion BTU/lb (gas) [1]	23,861	21,646	20,500
Heat of Combustion, (LNG) BTU/gal as stored (CNG) on vehicle (Gross) [2]	84,000 33,400	91,300	125,000
Density of Liquid @ 60°F, lb/gal	--	4.22	6
Density of Liquid @ -260°F, lb/gal	3.52	--	--
Density of gas @ 2000 psi, 60°F, lb/gal	1.4	--	--
Stoichiometric Air/ fuel Ratio			
lbs air/lb Gas	17.1	15.6	14.7
Ft^3 air/ft^3 Gas	9.52	23.82	--
Explosive Limits, % Vol in Air [3]			
Lower	5.3	2.37	1.3
Upper	13.9	9.5	6.0
Autoignition Temp [3]	999	871	495

[1] Perry's Chemical Engineers Handbook, Fourth Edition, pg. 3-142.

[2] Energy Density (see Table III for Energy Equivalents).

[3] Lange's Handbook of Chemistry, 14th Edition.

that natural gas can adequately be represented by
methane and LPG (liquefied petroleum gases) can
adequately be represented by propane. As these hydrocarbons usually constitute 95% of the respective fuels
in the marketplace, this simplification would not be
expected to introduce any significant bias or errors.
The values for gasoline are approximate in that there
are considerable variations among today's commercial
products.

The first items to be noted are the vapor
pressures and boiling points. Gasoline having a vapor
pressure of less than atmospheric is normally stored
on the vehicle in a light weight sheet metal tank sub-
jected to pressures usually not more than one pound
per square inch above or below atmospheric pressure.
Propane with a boiling point of -44°F can have a vapor
pressure exceeding 200 lbs. per square inch in a tank
standing in the hot sun. The ASME code for storage
tanks for this fuel require the tanks to be designed
for 312 pounds per square inch pressure while the
Federal Department of Transportation regulations
require that vehicle tanks be designed for a service
pressure of 240 pounds per square inch with a safety
factor of 4. Thus we have a requirement for a heavy,
high pressure tank. Further, you will note that the
heat of combustion of a gallon of LPG is substantially
less than the heat of combustion of gasoline.
Consequently, an LPG tank with a correspondingly larger
volume must be provided to permit the same vehicle
range or range must be sacrificed.

Turning to natural gas with a critical temperature
of -116°F, the tankage presents an even more difficult
problem. We have the choice of either carrying the
fuel as CNG (compressed natural gas) at very high
pressures, usually in industrial gas cylinders, at
pressures up to 2500 psi or carrying the material as
a cryogenic liquid. LNG (liquefied natural gas) has
a vapor pressure of one atmosphere at -259°F and a
vapor pressure of 45 atmosphere at its critical temperature of -116°F. Hence, a highly insulated and
expensive tank is required. Further, provision must
be made for safely venting boiled off fuel as the
temperature rises in a parked vehicle. These special
tanks need to be cylindrical because of strength con-

siderations and they need to be larger than the present
tank if equivalent cruising ranges are to be achieved.
These special tank shapes and sizes can not be
accommodated in the spaces usually allotted to the
normal gasoline tank. Hence, the gaseous fuel tank in
the most usual case usurps some of the otherwise useful
load space. Further, it is necessary to provide
special ventilation of this load space for safety
reasons if it is an enclosed space.

Figure 4 illustrates schematically the necessary
components of a retrofit LPG fuel system. From the
special fuel tank, the fuel passes first through an
automatic shut-off valve, required by law in most
states, thence through a fuel filter and a high
pressure regulator to a vaporizer where all of the
fuel is converted to the vapor state. The source of
heat for the vaporizer is usually the engine coolant
though there are instances where engine exhaust or
engine intake air has been used. From the vaporizer
the vapor passes through a second stage, low pressure
regulator and thence to a carburetor where it is mixed
with air and introduced into the engine manifold.
Hardware packages are available where the high pressure
regulator, evaporator and low pressure regulator are
all incorporated in a single unit referred to as a
converter. Gaseous fuel carburetors are available in
configurations which can be attached atop existing
gasoline carburetors where a dual fuel, gas or gasoline,
capability is desired for emergency or for long-dis-
tance travel. In the dual fuel instance there is
usually an electric interlock solenoid valve system to
prevent the flow of both fuels at the same time.
Natural gas fuel systems require the same functional
components except that the CNG system needs no evaporator. The design requirements of the components up to
the low pressure regulator are much more demanding in
the natural gas systems because of the very low
temperatures and/or very high pressures.

The retrofit modifications just described are
sufficient to convert a standard gasoline engine to
use gaseous fuels. Further, retrofit modifications
are possible and necessary if one wishes to optimize
the performance on the new fuel. Several of these
further modifications become quite expensive and are

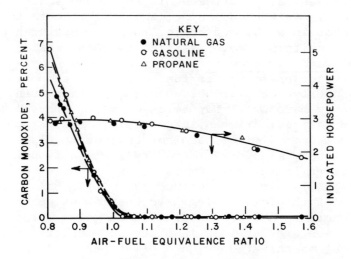

Figure 3. Carbon monoxide emissions and engine power as functions of air–fuel equivalence ratio at 50% throttle

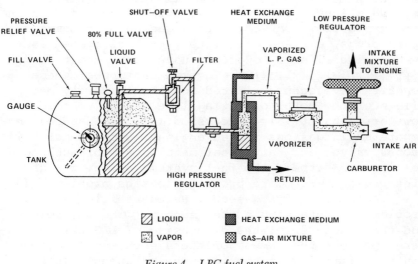

Figure 4. LPG fuel system

seldom employed. They are touched upon in some detail
in the reference (1) publication.

For vehicles which have been retrofitted both
performance benefits and shortcomings have been reported. Generally, driveability of gaseous fueled
vehicles is very good. Since the fuel reaches the
carburetor in a gaseous state, fuel/air mixtures can
be quite uniform. Cold starting is good, idling is
smooth, warm-up is quicker and there is no part
throttle surge if the air/fuel ratio adjustments have
been properly set. If the adjustments have been
improperly made or if NO_x emission control has been
over-emphasized, driveability problems similar to
those found in similar gasoline engines have resulted.
As previously mentioned there are frequently notice-
able losses in the maximum power and hence acceleration capability of the engine. (3), (4), (8a) Further,
because of the lower volumetric heating value of the
fuel, there is a substantial loss in miles per gallon
and an even greater loss in miles per tank filling for
the usual installation. On a miles per BTU basis,
LPG converted vehicles have been reported to give fuel
consumptions from 20% better (4) to 30% poorer (3)
than for gasoline. This wide range of results generally comes from comparing one optimized version against
one not similarly optimized on the alternate fuel.

On a longer term basis we again find benefits and
detriments from converting to a gaseous fuel system.
Spark plugs, oil and oil filter, exhaust system and
even engine life improvements have been reported. (4),
(6), (7), (8), (9), (10) In some instances, these
benefits have been estimated to save 1¼ to 1½¢ per
mile over the total operating life of the vehicle.
However, exhaust valve recession (4), (7) has been encountered in gaseous fueled engines which do not have
proper valves and valve seat metallurgy for such conversions. Further, the fuel systems being at high
pressure are much more prone to leakage problems. (4)

The driver of a gaseous fueled vehicle may have
problems in obtaining a fuel supply. The technology
and the hardware for the dispensing of LPG are well
developed but the number of the dispensing locations
may be a limiting factor in many instances. Thus
vehicles which can be most conveniently converted to

burn gaseous fuels would be fleet vehicles which operate on short trips from central terminals where a suitable refueling facility could be maintained as part of
the fleet service. Compressed natural gas is usually
handled under a pressure of about 2250 pounds per
square inch. This requires high pressure compressors,
tanks and lines. For LNG condensation by refrigeration
is required to yield a liquid product to fill the fuel
tank. Keeping this material in the liquid phase requires heavy insulation of all storage facilities.
While the technology for the compression or liquefac-
tion and dispensing of natural gas fuels is known,
hardware for this service may not be readily available
from off-the-shelf items.

The comparative safety of gasoline and the gaseous fuels should be mentioned. The relatively unknown, and thereby fearsome to the uninitiated, aspects
of handling fuel at high pressures and low temperatures
can be counteracted by adequate training and proper
hardware. The gaseous fuels have wider ranges of explosive mixture limits than does gasoline, thereby
possibly presenting a greater hazard in open fuel
spills. However, the faster rate at which these low
molecular weight fuels will dissipate in the atmosphere
tends to offset this wider explosive range problem.
In general, it can be said that both liquid and gaseous fuels can be handled safely if precautions in
installation, operation and maintenance of the fuel
system are taken. (4), (7), (11)

The relative costs of retrofit and gaseous fuel
operation are critical in determination of the
acceptability of retrofit in programs to the general
public. The cost of conversion of a vehicle to gaseous fuel depends primarily on the particular fuel and
secondarily on the required labor cost for installa -
tion and adjustment. When reasonable labor and overhead costs are used with the conversion equipment ex-
pensed in the current year the estimated cost for
conversion of each vehicle are $500 for an LPG system,
$600 for a CNG system and between $700 and $1,200 for
the LNG system. (1) The higher cost for the liquid
natural gas is attributable to the need for cryogenic
storage and the range of cost is due to the minimal
information in this area.   (All price estimates and

costs mentioned in this paper represent 1972 dollars
and material costs. These are now escalating rapidly
and at different rates for different items. Hence, in
any updating of the economics, the reader must pick
the costs applicable on a particular day on which he
hopes the estimates will be relevant.)

Fuel cost represents the most highly variable item
in published economic studies. Two rather comprehensive studies were selected for comparative analysis of
relative fuel economy. The United States General
Service Administration study (GSA), based on a twelve-
month, 24-vehicle comparison, (12) and the institute
of gas technology studies (IGT) (6) undertaken for the
EPA. Needless to say, the results of these two studies
do not agree with each other attributable primarily to
different basic assumptions affecting fuel cost. When
fuel costs were normalized using the same published
prices in both studies the differences between the two
were reduced but not totally eliminated. (1)

The GSA normalized results indicated compressed
natural gas to be a more economical fuel than gasoline
in Los Angeles, Houston, and Chicago but more expensive
in New York City. The IGT normalized results indicated
that none of the gaseous fuels was more economical than
gasoline. This difference in economics is due primarily to assumed compression costs at 33¢ per 1,000
standard cubic feet for large compressed natural gas
installation and 68¢ to 78¢ per 1,000 standard cubic
feet compression cost in small installations visualized
in the IGT report. In a number of economic studies
substantial fuel cost advantages were claimed for gaseous fuels but in most every instance this advantage
could be attributed to special tax concessions (13) or
other special situations which resulted in a lower
than normal cost for the gaseous fuel.

With the widely varying fluctuations in fuel
prices today, it would be futile to attempt to draw
any firm conclusions regarding relative fuel economy.
However, it is quite clear that there is no universal
economic advantage for any fuel for all vehicles.
Therefore, selection of the most economic fuel must be
based upon the unique economic position of fuel supply
and distribution as well as local and federal taxation
for each fleet operation.

In regard to sources and availability of gaseous fuels, little needs to be said at the present time to illustrate the current shortages of both natural gas LPG as well as gasoline. Let me summarize this aspect of the study with two statements.

1) Since supplies of natural gas and LPG already lag behind demand, any conversion of automotive type vehicles from gasoline to gaseous fuels can be accomplished only by withdrawing the gaseous fuel supplies from the current priority users.

2) New distribution and dispensing facilities would be required for greater utilization of LPG, CNG, or LNG, thereby requiring substantial capital investment by an industry facing critical requirements in this area to supply the gasoline needed in the near future.

In an effort to arrive at an estimate of the impact of such a retrofit program upon the automotive contribution of pollutants to the atmosphere we have first assumed that the average performance of a retrofitted vehicle would be equal to the average of the some 150 vehicles which have been retrofitted and for which emission measurements have been published. Figure 5 illustrates the individual emission factors used. The dotted lines represent the average emissions from the 150 retrofitted vehicles just mentioned. The average hydrocarbon emission level is 2.8 grams per mile. The average CO level is 12.2 grams per mile and the average NO_x level 3.5 grams per mile. The stepped solid lines represent the standards to which new vehicles are being built. Those of you who are familiar with recent EPA decisions will know that the last steps have been extended one year to 1976 for hydrocarbon and CO and to 1977 for NO_x. Meanwhile, there are some smaller steps due to new interim standards for 1975. The upper curved lines represent the weighted average emission rate for all gasoline vehicles in use. It will be noted from this chart that there is very little hydrocarbon incentive for retrofitting vehicles beyond the 1972 models. Starting with the 1975 models, the statutory limitation on hydrocarbon emissions will be actually below that achievable by gas retrofitted older vehicles. In the case of CO, there is some incentive for retrofit of all

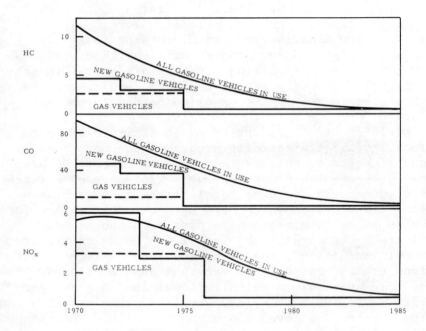

Figure 5. Emission factors for gasoline and gas powered vehicles. Passenger cars and light trucks—g/mile (1972 CVS basis).

vehicles up to those which meet the 1975 statutory
standards or even the 1975 interim standards which are
not plotted on this graph. With regard to NO_x there
is no incentive for gas retrofit beyond the 1973 models.
To arrive at the motor vehicle emission calculations
which are illustrated in the next figure, No. 6, it
has been necessary to make a number of further
assumptions.

    1. The retrofit program will be limited to high
mileage vehicles in fleets of 10 or more operating
largely in urban areas. It will be completed in 1975.

    2. The fuel consumption of the to be retrofitted
vehicles represents 12% of the national gasoline consumption.

    3. The gaseous fuels used for these retrofitted
fleets will be withdrawn from the industrial users
and replaced by 0.2% sulfur containing No. 2 distillate.
Demand for this distillate would as a result increase
by about 21% by 1975.

    The rationale for these assumptions is presented
in the full publication. (1)

    Figure 6 illustrates the impact on the total
vehicle emissions of utilization of gaseous fuels as
just described. The top of the lower line represents
the estimated reduction in emission rate from the
average gasoline powered motor vehicle assuming present
emission standards on new vehicles are adhered to. The
bottom of this lower line represents the average
emission rate for all motor vehicles if the hypothetical retrofit program outlined above were actually
implemented and completed by the end of 1975. The
thickness of this lower line is intended to represent
the benefit in terms of reduced emissions which would
be achieved as a result of this retrofit program.
At their maximum, these reductions would be a 3.9%
reduction in hydrocarbons, a 9.3% reduction in carbon
monoxide and a 1.6% reduction in nitrogen oxides. Due
to the phasing out of older vehicles these benefits
would be expected to substantially disappear in the
early 1980's.

    Figure 7 illustrates these reductions in another
way. On the right-hand side we see bars representing
the maximum reductions of emissions from the transportation sector to be achieved by this hypothetical

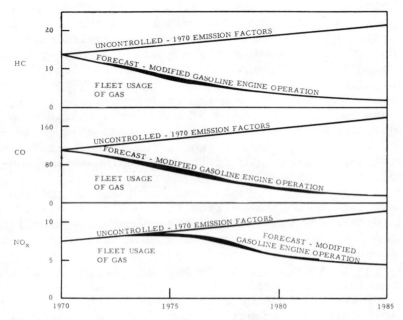

Figure 6. Motor vehicle emission improvements. Fleet use of gas in urban areas by 1975 (million tons/yr).

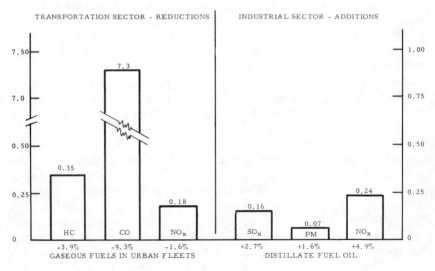

Figure 7. Net effect of gas-powered vehicles on pollutant emissions. Transportation and industrial energy sectors—1975 (million tons/yr.).

retrofit program. On the right hand side we see the increases in emissions by the industrial sector which would result if the industrial gaseous fuels were replaced by a 0.2% sulfur containing distillate. We would be trading off the hydrocarbon and carbon monoxide tonnages indicated for increases in sulfur dioxide and particulate matter. The net effect on nitrogen oxides would be nearly a stand off but even here some deterioration would be expected.

The net health effect of such changes in pollutant emission can be calculated on the basis of several published techniques. Using the three techniques summarized in the RECAT (14) report, we have calculated health effects ranging from a 2.1% improvement to a 2.5% deterioration. It seems doubtful if any of these numbers are meaningful.

Summarizing this assessment of the proposed retrofit program for substitution of gaseous fuels for gasoline, we would expect that:

1) Significant reductions of hydrocarbon, CO and NO_x emissions from older vehicles would result. However, these reductions would be small as a per cent of the emission of the total vehicle population and they would vanish by the early 1980's.

2) Utilization of gaseous fuels in the transportation sector would require substitution of heavier liquid fuel in the industrial sector resulting in increased industrial emissions of SO_2, NO_x and particulate.

3) A net environmental effect of the reductions of hydrocarbons and CO emissions and the increases in SO_2, NO_x and particulate anticipated as a result of such a program cannot be unequivocally established.

4) While the economics of certain existing gaseous fuel retrofits appear attractive, these usually involve fuel tax concessions or other special circumstances resulting in low fuel costs.

5) In any mandatory retrofit program expanding the use of gaseous fuels, the economy would have to bear the costs of new fuel supply and dispensing facilities, retrofitting costs for both vehicles and for industries from which the gaseous fuels have been withdrawn and the rebalancing of refinery facilities to meet the changed product slate.

Literature Cited

1. Gaseous Motor Fuels - An Assessment of the Current
and Future States -- American Petroleum Institute,
Committee on Environmental Affairs, API Publication
No. 4186, August (1973).
2. Allsup, J. R. and Fleming, R. D., "Emission
Characteristics of Propane as Automotive Fuel",
Bartlesville Energy Research Center, Bureau of Mines
Report of Investigations 7672, (1972).
3. Genslak, S. L., "Evaluation of Gaseous Fuels for
Automobiles," SAE Paper 720125, January (1972).
4. Gaseous Fueled Vehicles and the Environment, GSA
Symposium, May 24-26, (1972).
5. Baxter, M. C., "How Does LP Gas Reduce Exhaust
Emissions," The Compressed Gas Industry and Its
Relationship to the Environment, 58th Annual Meeting,
January (1971).
6. Emission Reduction Using Gaseous Fuels for Vehicular
Propulsion, Institute of Gas Technology, June (1971),
PB-201410.
7. Bintz, L. J., Kramer, M., and Tappenden, T. A.,
"LP-Gas and the Automobile Club of Southern California,"
ASTM-NGPA, NLPGA Symposium on LP-Gas Engine Fuels,
Los Angeles, California, June 27, (1972).
8. McJones, R. W., "Method and System for Reducing
Oxides of Nitrogen and Other Pollutants from Internal
Combustion Engines," U. S. Patent 3,650,255, March 21,
(1972).
8a. Environmental Protection Agency, Position Paper -
Conversion of Motor Vehicles to Gaseous Fuel to Reduce
Air Pollution, April (1972).
9. Douglas, L. A., "Dual Fuel Systems Design - CNG/
Gasoline," The Compressed Gas Industry and Its Rela-
tionship to the Environment, 58th Annual Meeting,
January (1971).
10. Diamond, R., "LPG A Now Answer to Air Pollution,"
SAE Automotive Engineering Exposition, January (1971).
11. California Institute of Technology, Environmental
Laboratory, "Caltech Clean Air Car Project -- Gaseous
Fuels Manual," March 1, (1972).
12. Pollution Reduction with Cost Savings, U. S.
Government General Services Administration, GSA DC 71-
10828.

13. Sagan, V. R., "Current and Pending Legislation for Gaseous Fuels," ASTM-NGPA-NLGPA Symposium on LP-Gas Engine Fuels, Los Angeles, California, April (1972).

This paper is a summary of API publication No. 4186 with the same title and authors for presentation to the Symposium on Current Approaches to Automotive Emission Control - ACS Meeting, Los Angeles, April 1, 1974. The full publication is available from the American Petroleum Institute, 1801 K Street, N.W., Washington, D.C. 20006.

4

Fuel Volatility as an Adjunct to Auto Emission Control

R. W. HURN, B. H. ECCLESTON, and D. B. ECCLESTON

Bartlesville Energy Research Center, Bureau of Mines, Bartlesville, Okla. 74003

Abstract

Late-model vehicles were used in an experimental study of
the interaction of fuel volatility with emissions and associated
fuel economy. Volatility characteristics of the test fuels
ranged between 7 and 14 pounds Reid vapor pressure; 50 pct point
between 130° and 240° F; and 90 pct point between 190° and 370° F.
Choke settings of each vehicle were adjusted as needed for choke
action appropriate to each fuel's volatility.

Midrange and back-end volatility were found to influence
emissions but in a possibly complex interaction of the two. The
principal influence is upon emissions during cold start and
warmup. Within the vapor pressure limits traditional of U.S.
fuels, vapor pressure and fuel front-end volatility were found
to have only slight effect upon either emissions or fuel economy.

Results show that in this study, reduced hydrocarbon and
carbon monoxide emissions are associated with the fuels of lower
back-end volatility. Among the fuels tested higher fuel economy
values were found with the heavier fuels. However, the differences in fuel economy may--and probably do--result from differences in mixture stoichiometry and fuel heating value rather
than from differences in fuel volatility, per se.

Introduction

In 1967, the Bureau of Mines, in cooperation with the American Petroleum Institute, began a series of experiments to obtain
more comprehensive information on the effects of fuel volatility
on exhaust emissions. Results of the API-sponsored studies
showed that fuel volatility does affect exhaust emissions but
only in a complex relationship in which vapor pressure mid-range
and back-end volatility interact. Briefly, it was concluded from
the 1967-71 studies that, for the 1966-69 model year cars that
were tested, and for operation at moderate ambient temperatures--

1. Exhaust hydrocarbon (HC) and carbon monoxide (CO) were
   increased when vapor pressure was decreased below 9 pounds
   Reid vapor pressure (RVP), and
2. HC and CO generally were decreased by increasing mid-range
   volatility as compared with typical current practice.

   With this work having established evidence of a strong
relationship between fuel volatility and exhaust emissions, the
Bureau conducted follow-on studies to delineate more clearly the
effects of mid-range and back-end volatility. The work sought
answers to two questions in particular:

1. If fuel volatility is increased, what is the likely effect
   on emissions and fuel economy for vehicles currently on the
   road?
2. If fuel volatility is increased, and with minor modifications
   to carburetion, can emissions be held the same or lowered
   while increasing fuel economy?

   To provide some answers to these questions, experiments were
designed to use 12 fuels in 3 cars: (1) A compact car with a
250-CID engine and single-barrel carburetor, (2) a light sedan
with 350-CID engine and 2-bbl carburetor, and (3) an intermediate
sedan with 350-CID engine and 4-bbl carburetor. All were equipped
with automatic transmission and with power options typically sold
with that class of vehicle.

   The fuels were designed to represent, insofar as possible,
options for volatility change selectively within the front, mid-
range, or back-end portion of the fuel boiling range. Combinations of these selective changes also were designed into the
fuels.

   Emissions were measured using analytical instruments and the
driving cycle as specified for the 1975 Federal test procedure.
All work was done at controlled temperature and humidity; data
were taken at test temperatures of 20°, 55°, and 90° F. This is
a summary report of findings, and except for one reference to
test data taken at 90° F, references are limited to the data
obtained at 55° F as being illustrative of trends that were
represented generally.

Discussion

   Turning now to a more detailed discussion of the study, let
me first review briefly the volatility characteristics of the
12 test fuels. Vapor pressures of the fuels ranged roughly
between $7\frac{1}{2}$ and 14 pounds. Higher vapor pressure fuels were
excluded from the 90° F test; the lower vapor pressure fuels
were excluded from tests made at 20° F. The 10 pct boiling
points of the fuels ranged from about 100° F in one of the more
volatile fuels to 146° F in one of the least volatile. The
50 pct points of the fuels ranged between about 130° and 250° F
among the 12 fuels; the 90 pct point was varied between roughly

185° and 375° F. As indicated earlier, various combinations of
these fuel characteristics were used in designing the 12 fuels.
The range of characteristics is summarized below.

Test Fuels Volatility Characteristics

	Range of Values
Vapor Pressure, RVP, lb.	7.4 - 14.0
10 pct point, °F	100.0 - 146.0
50 pct point, °F	130.0 - 254.0
90 pct point, °F	184.0 - 376.0

Total of 12 fuels

A detailed inspection of the emissions data showed unmistakable differences attributable to fuels. No one volatility
characteristic (i.e., RVP, 10 pct point, 50 pct point, or 90
pct point) was shown to act alone in a dominant fashion.
However, the combination of the 50 pct and 90 pct boiling points,
a measure of volatility across the full back half, does permit
correlation with emissions. At least the combination appears
to exert a greater influence than any other one volatility
characteristic or any other combination of vapor pressure, and
10, 50, or 90 pct boiling point. In this paper we have, therefore, elected to show the relationship that was found to exist
between emissions and the volatility of the fuel back-end as
expressed in the sum of the 50 pct and 90 pct boiling points.
In figure 1 and in following figures, volatility as expressed
by the sum of the 50 pct and 90 pct boiling points will be shown
as a horizontal variable. The observed emissions are shown as
the vertical coordinate. Also, in figure 1 as in those following,
the data points are shown for one vehicle, a compact car. The
trend of data for the other cars, the intermediate and the light
sedan, are shown by the dashed curves.
These data reflect a general trend, but although the trend
is real, there is nonetheless only a loose relationship between
the volatility characteristics and exhaust emissions. If the
data in figure 1 were the only measurements available, they would
not be significant in indicating the trends. However, as is to
be seen in the figures to follow, other measurements also indicate
trend lines, and taken together there is a high degree of confidence in the trends that are shown. Therefore, we conclude that
HC emissions tend to be somewhat increased with back-end volatility either increased or decreased from the intermediate range of
the volatility characteristic.

Oxides of nitrogen (NO_x) also is shown to be responsive to the fuel back-end volatility characteristic (Fig. 2). The data are internally self-consistent and consistent between automobiles, and as earlier indicated, we believe that the trends exist as shown. The explanation for increased NO_x with the lower volatility fuels probably relates not to volatility characteristic, per se, but rather to the stoichiometry of the air-fuel mixture and to heat content of the fuel. This point is explored further in discussions to follow.

Carbon monoxide data (Fig. 3) follow the same general trends as shown for hydrocarbon; however, the effect does appear to be somewhat more pronounced in showing CO increased with the fuels that have 50 pct plus 90 pct point values below the 450° to 500° F range. Also consistent with the hydrocarbon data, CO emissions were found to increase with back-end volatility decreased well beyond the intermediate range of the fuels tested. In general, trends for the different emissions are mutually consistent in that CO and HC move together and increased NO_x emissions are associated with decreased CO and HC.

The effect of increased emissions with fuels of distinctly heavier back-ends is shown accentuated in the 90° F data for the light sedan (Fig. 4). In the case of these data, if the two deviated points "A" and "B" are not considered, emissions are shown to increase in a consistent pattern. However, if the back-half volatility is characterized by low mid-point volatility, but combined with high back-end volatility, point "A", emissions are increased over that of a fuel typically more representative. If, on the other hand, the same average volatility characteristics are achieved by combining high mid-range with low back-end volatility, point "B", CO emission is decreased. This effect appeared to be consistent with the three vehicles; therefore, we conclude that emissions reduction is best served by incorporating high mid-range volatility in the fuel.

One question to be answered from results of the tests was whether minor modifications to the carburetion system--to be made in connection with increased volatility--could be used advantageously as an emissions control measure. To obtain information on this point, emissions were measured on the vehicles both as set to standard specifications and with the choke schedule modified to effect less choking and more rapid choke release. The results were, in all cases, to realize marked reduction in emissions during cold starts by modifying the choking schedule toward less choking. Using a standard choke installation, CO emissions were increased by about 4 grams per mile (over the emissions measured with the reference fuel) by increasing fuel volatility at the 10 pct point. However, with a modified choke schedule (and still achieving satisfactory vehicle operation), CO emissions from the fuel with volatility increased at the 10 pct point decreased by about 4 grams per mile from the level measured with the reference fuel. In like manner,

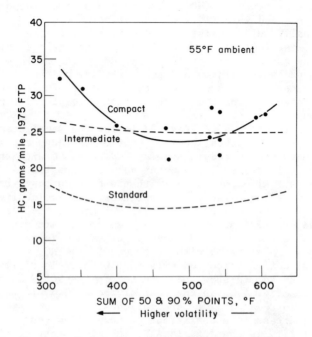

Figure 1. *Hydrocarbon emissions response to mid-range and back-end volatility*

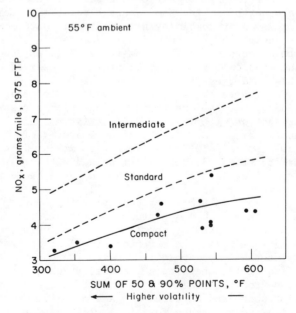

Figure 2. *NO$_x$ response to mid-range and back-end volatility*

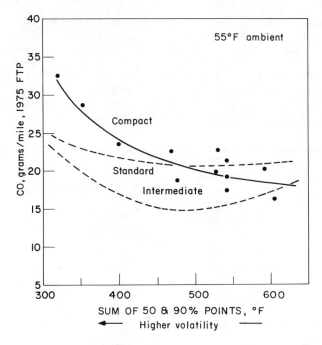

Figure 3. Carbon monoxide response to mid-range and back-end volatility

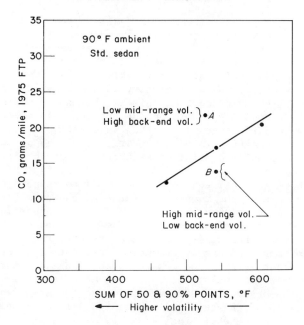

Figure 4. Carbon monoxide response to mid-range and back-end volatility

significant reduction in CO emissions was achieved using a
modified choke schedule when fuel volatility was increased at
the 50 pct point, and also when volatility was increased throughout the back half of the boiling curve. However, consistent
with earlier evidence, increasing volatility at the 90 pct point
reduced emissions less significantly than did the other fuel
modifications just cited. These relationships of volatility
adjustments to emissions are summarized below:

Effect of Volatility Change in CO Emissions--55° F
Ambient Cold Start

Fuel Modification	Effect on CO Emissions (gm/mile)	
	Standard Choke Sch.	Modified Choke Sch.
Volatility increased:		
at 10 pct point	+4.2	- 3.8
at 50 pct point	-3.0	-17.4
at 90 pct point	+4.6	- 2.0
throughout midrange	+9.8	- 9.8

    Fuel economy was determined for all of the tests. Results
showed unmistakable response to fuel change. As is to be
observed from the curves of figure 5, fuel economy improved
measurably with the heavier fuels and the effect was consistent
among the three vehicles. Conceivably, the change in economy
could be associated with volatility characteristics, per se;
but there is no evidence that this is the case. Quite the
contrary, it would appear to be due to change in air-fuel
ratio and to differences in heating value between the fuels.
    When associated with mixture ratio, fuel economy was found
to be directionally consistent with the change in air-fuel ratio
that was measured for each of the tests (Fig. 6). More significant, however, fuel economy is believed to respond to the
energy content of the respective fuels as shown estimated in
figure 7. As a measure of the rough correspondence found
between fuel economy and heating value, it is noted that fuel
economy between the worst and best cases differed by about 11
pct. The corresponding difference in heating value of the
fuels is about 17 pct. We conclude, therefore, that there is no
evidence that fuel volatility, per se, affected fuel economy,
but that the observed differences are due to variations in
heating value of the fuel and associated influence upon the
stoichiometry of the air-fuel mixture.

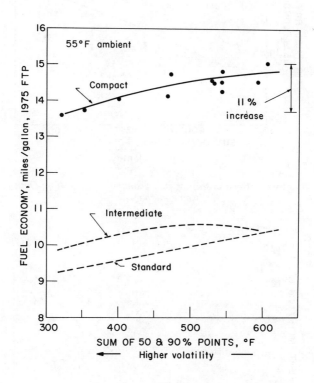

Figure 5.   Fuel economy at 55 F ambient

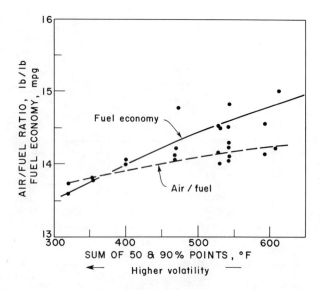

*Figure 6.   Relationship of fuel economy to volatility. Data
from compact car.*

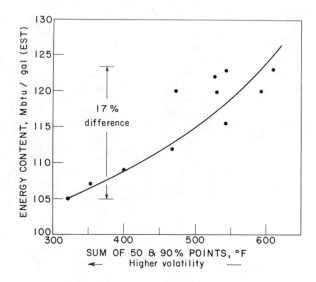

Figure 7.   Energy content of test fuels

5

Pre–engine Converter

N. Y. CHEN and S. J. LUCKI

Mobil Research and Development Corp., Princeton 08540 and Paulsboro 08066, N. J.

I. Introduction

During the past decade, efforts to reduce vehicular pollutant emission have included suggestions for
the removal of lead from gasoline or for use of alternative fuels such as H_2 and low molecular weight hydrocarbons which are known to have high octane values
and good burning characteristics (1), (2), (3). Lead
removal, which is already being implemented, raises
the octane requirement of the fuel. The increased
severity required in refining to produce such high
octane gasoline decreases the gasoline yield per bar-
rel of crude and, therefore, increases crude oil consumption and demands more refining capacity in the
face of an impending energy crisis. Use of low molecular weight hydrocarbons is difficult to implement
because of safety hazards and lack of nationwide storage and distribution systems.

The concept of attaching a catalytic reactor to
an internal combustion engine converting liquid hydrocarbons to gaseous fuel was disclosed in a patent
issued to Cook (4) in 1940. Recently, it has received
some renewed attention. Newkirk et al (2), (3) described their concept of an on-board production of
CO_2/H_2 mixture by steam reforming of gasoline fuel. A
U.S. patent was issued to W. R. Grace Company in 1972
(5) on a mobile catalytic cracking unit in conjunction
with a mobile internal combustion engine. In 1973,
Siemens Company (6) of Germany announced a "splitting
carburetor" which breaks up gasoline and related fuels
into burnable gases. The jet propulsion laboratory
of NASA (7) is investigating the concept of the generation of hydrogen for use as an additive to gasoline
in internal combustion engines.

While little technical data are available, these

developments appear to represent different approaches
of adapting established industrial catalytic processes
designed for a narrow range of operating conditions to
moving vehicles which must operate from idling to full
throttle.

To design a reactor system capable of operating
satisfactorily under full throttle conditions requires
either a large reactor or an unusually active and
efficient catalyst. A standard 300 cu. in. automotive
engine at full throttle consumes fuel at a rate of
about 10 cc/sec. If the reactor were operating at 1-2
LHSV (vol/vol/hr.), i.e., at the throughput of an
average industrial reactor, the engine would require
a catalyst bed volume of 18-36 liters (4.8 - 9.5
gallons) - far larger than the carburetor it replaces.
The necessity of a multi-reactor system for continuous
operation plus accessory devices including the fuel
preheater, etc., would make the system impractically
bulky and too slow to warm up. Therefore, a workable
system clearly depended on the discovery of new catalysts of high activity. To reduce the size of the
reactor to that of a carburetor, an increase in catalytic activity by a factor of at least 50-100 is
necessary.

In addition to the problem of the catalytic reactor volume, the life of the catalyst is also of critical importance. Most industrial catalysts require
periodic oxidative regeneration in a matter of minutes
after operation to maintain their effectiveness. A
catalytic cracking catalyst, such as that proposed in
the patent issued to Grace (5), requires frequent
regeneration. An example described in this patent
states that with a zeolite-containing catalyst, 2% of
the fuel was converted to coke in the catalytic converter. We estimate that under the proposed operating
conditions, each volume of catalyst could process no
more than 10 volumes of fuel before sufficient coke
(20%) would have deposited on the catalyst to deacti-
vate it. Thus, even if the catalyst were active
enough to operate at 100 LHSV, no more than 6 minutes
of continuous operation between oxidative regenera-
tions would be feasible.

In this paper we present experimental studies
with a catalyst that overcomes many of these limitations.

II. Experimental

1. Equipment. To demonstrate the performance of
the new catalyst system, the catalytic reactor was

attached to a standard motor knock Test Engine, Method IP44/60 (8), bypassing the carburetor. Figure 1 shows a schematic diagram of the catalytic unit. The reactor consisted of two 3/4 inch O.D. x 18 inches stainless steel cylinders mounted vertically, one on top of the other, and connected in series. The top chamber serving as the preheater contained 82 cc of 3 mm diameter glass beads; the catalyst bed (5 inches long) in the bottom chamber consisted of 24 cc of 20/30 mesh catalyst mixed with 12 cc of 8/14 mesh Vycor chips. During thermal runs, Vycor chips were substituted for the catalyst.

Both cylinders were electrically heated. Liquid fuel flow was metered with a rotameter. Air-fuel ratio was monitored by Orsat analysis of the exhaust.

2. <u>Test Fuels</u>. Two types of feedstocks were used: (1) an 86 research octane (R+O) and 79.5 motor octane (M+O) reformate obtained from Mobil's Paulsboro Refinery containing: 23.4 wt. % n-paraffins, 33.9% branched paraffins, 1.2% olefins, 1.0% naphthenes and 40.5% aromatics, and (2) a Kuwait naphtha of 40.5 clear motor octane (M+O).

III.   Results and Discussions

1. <u>Upgrading of a C$_5$-400°F reformate</u>. The experiments were carried out by charging the liquid fuel stored in a pressurized reservoir (4500 cc) at 38 cc/min. continuously for about 2 hours through the catalytic converter during which time the motor octane number of the reactor effluent was determined every 30 minutes. At the end of 2 hours, the reactor was cooled to 800°F with purge nitrogen while the fuel reservoir was being refilled. The experiment was then repeated. Two catalysts were examined, viz., a new stable zeolite catalyst (12 gms) and a commercial zeolite cracking catalyst (16 gms), which had previously been aged for 2 hours in a test described later. The feed rate corresponds to a weight hourly space velocity of 140 and 93, respectively. The reactor was maintained at between 910 and 920°F. Octane rating of the reactor effluent as a function of the cumulative on-stream time is shown in Figure 2. During the first 2 hours, the stable zeolite raised the octane number from the thermal value of 79.6 to 85 M+O. The octane dropped 2 numbers during the next two hours and maintained above 82 M+O for the next seven hours. The aged commercial zeolite catalyst, on the other hand, produced no appreciable conversion

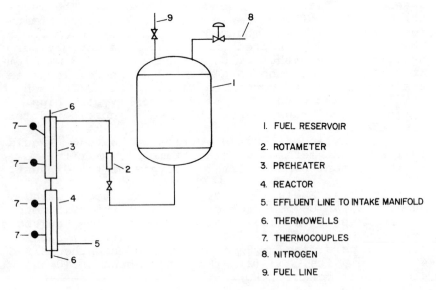

1. FUEL RESERVOIR

2. ROTAMETER

3. PREHEATER

4. REACTOR

5. EFFLUENT LINE TO INTAKE MANIFOLD

6. THERMOWELLS

7. THERMOCOUPLES

8. NITROGEN

9. FUEL LINE

Figure 1.   Diagram of the catalytic unit

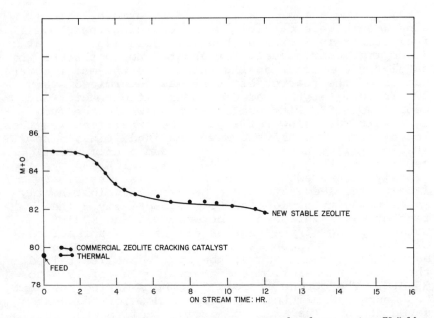

Figure 2.   Effect of on-stream time on octane rating—knock test engine. 79.5 M + O, C_5–400 reformate, 915°F, 95 LHSV.

under the experimental conditions, i.e., at this high
space velocity.

2. Upgrading of a C_6-350 Kuwait naphtha. After
12 hours of operation without regeneration using reformate as the feed, the fuel was switched to the low
octane virgin naphtha and the test continued over the
aged stable zeolite catalyst for an additional two
2-hours runs before the experiment was terminated due
to a mechanical malfunction. The result of the naphtha test is summarized in Figure 3. Shown in the same
figure are the results over a fresh commercial catalyst and the thermal run. It is interesting to note
that a boost of 22 motor octane numbers was registered
by the aged stable zeolite catalyst while the fresh
commercial zeolite cracking catalyst and the thermal
run recorded a gain of only 10 and 6 numbers, respectively.

3. Shape Selective Cracking. In addition to
their excellent burning qualities, i.e., non-polluting
combustion, light hydrocarbons have volume blending
octane ratings ranged between 100 and 150 research
clear numbers (R+O). Thus low octane liquids such as
virgin naphtha and mildly reformed reformate can be
upgraded by partially converting them to light hydrocarbons in the pre-engine converter, and feeding the
entire reactor effluent directly into the engine.
A typical distribution of reaction products is
shown in Table I for three samples collected when a
blend of C_6 hydrocarbons was passed over the stable
zeolite catalyst at 1 atm. and 900°F. The results
clearly show that the catalyst exhibited preferential
shape selective cracking in the order of n- > mono-
methyl- > dimethyl-paraffins. Thus isomers having the
lower octane ratings are preferentially cracked. The
C_4 minus cracked products are highly olefinic and some
C_7^+ aromatics are formed by secondary reactions.
The added advantage of shape selective cracking
in the order of octane rating is illustrated by the
cracking of a 61 research octane Udex raffinate, a low
octane product from the solvent extraction of aromatics from a reformate. In Figure 4, curve 1 shows the
calculated octane number of the reactor effluent vs.
wt. % liquid cracked to C_4^- light hydrocarbons. To
produce a 91 R+O fuel, about 49% of the liquid is
cracked. Examination of the composition of the raffinate shows that the straight chain paraffins having
an average octane rating of 17 R+O represent 27% of
the liquid. Curve II shows that when these

TABLE I

Product Distribution at 900°F

Wt. %	Feed	WHSV			% Conversion		
		55	100	200	55	100	200
Methane	--	0.6	0.4	0.2			
Ethane, Ethene	--	3.3	2.5	0.8			
Propane	--	9.7	6.1	1.1			
Propene	--	7.0	5.8	3.6			
Butanes	--	2.1	1.5	0.3			
Butenes	--	3.1	2.6	2.0			
2,2-Dimethylbutane	9.1	8.6	8.6	8.6	5.5	5.5	5.5
2,3-Dimethylbutane	5.4	5.4	5.4	5.4	0.0	0.0	0.0
2-Methylpentane	13.5	8.5	10.7	11.7	37.0	20.7	13.3
Hexane, 1-hexene	24.5	6.9	10.4	16.1	71.8	57.6	34.3
Benzene	47.5	38.1	40.8	44.4	19.8	14.1	6.5
C_7^+ Aromatics	--	6.7	5.1	5.6			
R+O	77.1	96.8	92.0	84.5			
ΔON	--	19.7	14.9	7.4			
C_4^- % Conversion	--	25.8	18.9	8.0			
ΔON/% Conversion	--	0.76	0.79	0.93			

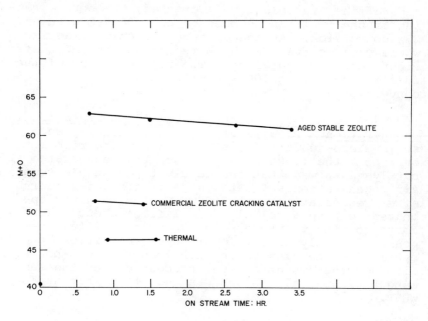

Figure 3.   Effect of on-stream time on octane rating. 405 M + O, Kuwait naph-tha, 915°F, 95 LHSV.

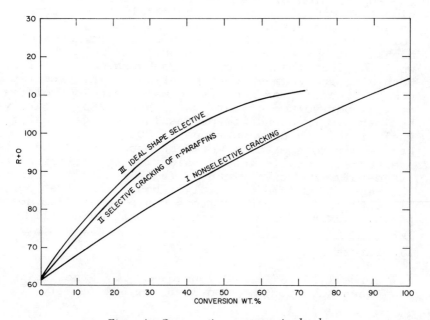

Figure 4.   Octane rating vs. conversion level

n-paraffins are selectively cracked, the octane rating of the fuel can be boosted to ~ 90 R+O with only 27% conversion. An ideal shape selective cracking would yield curve III which represents the most efficient route of upgrading. The octane rating of the fuel is boosted to 100 R+O with less than 40% conversion.

4. Catalyst life and stability toward oxidative regeneration. Preliminary data obtained in bench scale micro-reactor (9) studies using the reformate over both the fresh catalysts and the regenerated catalysts showed that the catalyst was stable toward air regeneration and catalyst activity was restored after regeneration. At 100 WHSV, the catalyst appeared to have a useful cycle life of about 7 hours, corresponding to processing 700 pounds of fuel per pound of catalyst. At lower space velocities, the cycle life appeared to be much longer than 7 hours, although the amount of fuel processed over the same length of operating hours was less than that at 100 WHSV.

IV. Conclusions

A highly active, stable and shape selective zeolite cracking catalyst overcomes a major problem in the application of the concept of pre-engine conversion to a moving vehicle. The catalyst is active enough to operate at above 100 LHSV and 900°F. The volume of a catalyst bed for a 300 cu. in. engine capable of operating satisfactorily at full throttle would be less than 360 cc - a manageable volume from both size and warm-up considerations. The catalyst has the capacity of processing more than 700 volumes of fuel per volume of catalyst. For a 360 cc catalyst bed, this corresponds to processing 66.5 gallons of fuel or about a driving range of 800 to 1000 miles before air regeneration would be necessary. The catalyst appears stable toward oxidative regeneration and its catalytic activity can be fully restored. Since the required volume of catalyst bed is small enough, segmented or multiple reactors could be used to accomplish cracking operation and regeneration at all times.

Literature Cited

1. Corbeil, R. J. and Griswold, S. S., Proc. Int. Clean Air. Congr., 2nd 1970, p. 624 (1971).

2. Newkirk, M. S. and Abel, J. L., paper No. 720670 presented at New England Section Meeting, Society of Automotive Engineers, November 2, 1971.
3. U. S. Patent 3,682,142 assigned to International Materials Corp., August 8, 1972.
4. U. S. Patent 2,201,965 to John T. Cook, May 21, 1940.
5. U. S. Patent 3,635,200 assigned to W. R. Grace and Company, January 18, 1972.
6. New York Times, February 11, 1973.
7. New York Times, September 17, 1973.
8. IP Standard for Petroleum and its Products, Part II, 2nd Ed. Inst. Petrol., London, 1960.
9. Chen, N. Y. and Lucki, S. J., Ind. Eng. Chem. Process Des. Develop. (1971), 10, 71.

6

Low Emissions Combustion Engines for Motor Vehicles

HENRY K. NEWHALL

Chevron Research Co., Richmond, Calif. 94802

During the past 10-15 years, very significant advances in controlling exhaust emissions from automobile power plants have been made. Initially, emissions reductions were achieved through careful readjustment and control of engine operating conditions (1). More recently, highly effective exhaust treatment devices requiring a minimum of basic modification to the already highly developed internal combustion engine have been demonstrated. These are based on thermal and/or catalytic oxidation of hydrocarbons (HC) and carbon monoxide (CO) in the engine exhaust system (2,3,4). Nitrogen oxide (NO_x) emissions have been reduced to some extent through a combination of retarded ignition timing and exhaust gas recirculation (EGR), both factors serving to diminish severity of the combustion process temperature-time history without substantially altering design of the basic engine (5).

Basic combustion process modification as an alternative means for emissions control has received less attention than the foregoing techniques, though it has been demonstrated that certain modified combustion systems can in principle yield significant pollutant reductions without need for exhaust treatment devices external to the engine. Additionally, it has been demonstrated that when compared with conventional engines controlled to low emissions levels, modified combustion processes can offer improved fuel economy.

Nearly all such modifications involve engine designs permitting combustion of fuel-air mixtures lean beyond normal ignition limits. As will be shown, decreased mean combustion temperatures associated with extremely lean combustion tend to limit the rate of nitric oxide (NO) formation and, hence, the emission of NO_x. At the same time, the relatively high oxygen content of lean mixture combustion products tends to promote complete oxidation of unburned HC and CO provided that combustion gas temperatures are sufficiently high during late portions of the engine cycle.

The purpose of this paper is to present an overall review of the underlying concepts and current status of unconventional engines employing modified combustion as a means for emissions control. Detailed findings related to specific power plants or to specific applications will be treated by the papers which follow.

Throughout the paper exhaust emissions will be compared with emissions standards legislated for the years 1975 and 1976. As a result of Environmental Protection Agency (EPA) actions suspending the 1975 HC and CO standards and the 1976 NO_x standard, several sets of values exist. These are listed in Table I and in the text

Table I

Federal Exhaust Emissions
Standards, Grams/Mile[1]

	1975			1976			1977
	Statutory	U.S.	Interim California	Statutory	U.S.	Interim California	Statutory
HC	0.41	1.5	0.9	0.41	0.41	0.41	0.41
CO	3.4	15	9.0	3.4	3.4	3.4	3.4
NO_x	3.0	3.1	2.0	0.4	2.0	2.0	0.4

[1]As measured using 1975 CVS C-H procedure.

will be referenced either as statutory (original standards as set by the Clean Air Act Amendment of 1970) or as interim standards as set by the EPA.

Theoretical Basis for
Combustion Modification

Figure 1 has been derived from experimental measurements (6) of the rate of NO formation in combustion processes under conditions typical of engine operation. This figure demonstrates two major points related to control of NO_x emissions: first, the slow rate of NO formation relative to the rates of major combustion reactions responsible for heat release and, second, the strong influence of fuel-air equivalence ratio on the rate of NO formation.

Experimental combustion studies (7) employing "well-stirred reactors" have shown that hydrocarbon-air combustion rates can be correlated by an expression of the form

$$\frac{N}{V \, p^{1.2}} = 48 \, \frac{Gram\text{-}Moles/Liter\text{-}Second}{Atm^{1.8}}$$

where:

> N = moles reactants consumed per second
> V = combustion volume
> p = total pressure

For conditions typical of engine operation, this expression yields a time of approximately 0.1 ms for completion of major heat release reactions following ignition of a localized parcel of fuel-air mixture within the combustion chamber. Comparison with Figure 1 shows that the time required for formation of significant amounts of NO in combustion gases is at least a factor of 10 greater. Thus, in principle, energy conversion can be effected in times much shorter than required for NO formation. In the conventional spark ignition engine, the relatively lengthy flame travel process permits combustion products to remain at high temperatures sufficiently long that considerable NO formation occurs.

Figure 2, which consolidates the data of Figure 1, indicates that maximum rates of NO formation are observed at fuel-air equivalence ratios around 0.9 (fuel lean). For richer mixtures, the concentrations of atomic and diatomic oxygen, which participate as reactants in the formation of NO in combustion gases, decrease. On the other hand, for mixtures leaner than approximately 0.9 equivalence ratio, decreasing combustion temperatures result in lower NO formation rates.

Figure 2 serves as a basis for combustion process modification. Operation with extremely rich fuel-air mixtures (Point A of Figure 2), of course, results in low NO_x emissions since the maximum chemical equilibrium NO level is greatly reduced under such conditions. However, the resultant penalties in terms of impaired fuel economy and excessive HC and CO emissions are well known. An alternative is operation with extremely lean mixtures (Point B) - lean beyond normal ignition limits. Combustion under such conditions can lead to low NO_x emissions while at the same time providing an excess of oxygen for complete combustion of CO and HC.

Operation of internal combustion engines with extremely lean overall fuel-air ratios has been achieved in several ways, employing a number of differing combustion chamber configurations. One approach involves ignition of a very small and localized quantity of fuel-rich and ingitable mixture (Point A of Figure 2), which in turn serves to inflame a much larger quantity of surrounding fuel-air mixture too lean for ignition under normal circumstances. The bulk or average fuel-air ratio for the process corresponds to Point B of Figure 2; and, as a consequence, reduced exhaust emissions should result.

A second approach involves timed staging of the combustion process. An initial rich mixture stage in which major combustion reactions are carried out is followed by extremely rapid mixing of rich mixture combustion products with dilution air. The

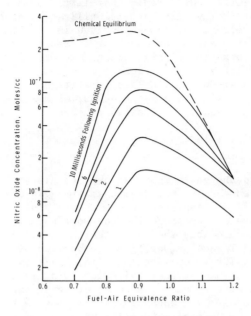

*Figure 1. Rate of nitric oxide formation in
engine combustion gases (6)*

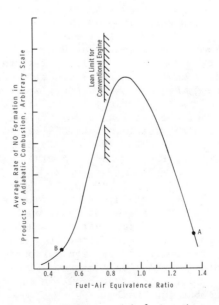

*Figure 2. Influence of fuel–air ratio on
nitric oxide formation rate*

transition from initial Point A to final Point B in Figure 2 is, in principle, sufficiently rapid that little opportunity for NO formation exists. Implicit here is utilization of the concept that the heat release reactions involved in the transition from Point A to Point B can be carried out so rapidly that time is not available for formation of significant amounts of NO.

Reciprocating spark ignition engines designed to exploit the foregoing ideas are usually called stratified charge engines, a term generally applied to a large number of designs encompassing a wide spectrum of basic combustion processes.

Open-Chamber Stratified Charge Engines

Stratified charge engines can be conveniently divided into two types: open-chamber and dual-chamber engines. The open-chamber stratified charge engine has a long history of research interest. Those engines reaching the most advances stages of development are probably the Ford-programmed combustion process (PROCO) (8) and Texaco's controlled combustion process (TCCS) (9). Both engines employ a combination of inlet air swirl and direct timed combustion chamber fuel injection to achieve a local fuel-rich ignitable mixture near the point of ignition. The overall mixture ratio under most operating conditions is fuel lean.

The Texaco TCCS engine is illustrated schematically in Figure 3. During the engine inlet stroke, an unthrottled supply of air enters the cylinder through an inlet port oriented to promote a specified level of air swirl within the cylinder and combustion chamber. As the subsequent compression stroke nears completion, fuel is injected into and mixes with an element of swirling air charge. This initial fuel-air mixture is spark ignited, and a flame zone is established downstream from the nozzle. As injection continues, fuel-air mixture is continuously swept into the flame zone. The total quantity of fuel consumed per cycle and, hence, engine power output, are controlled by varying the duration of fuel injection. Under nearly all engine operating conditions, the total quantity of fuel injected is on the lean side of stoichiometric. The TCCS system has been under development by Texaco since the 1940's. To date, this work has involved application of the process to a wide variety of engine configurations.

Like the TCCS engine, the PROCO system (Figure 4) employs timed combustion chamber fuel injection. However, in contrast to the TCCS system, the PROCO system is based on formation of a pre-mixed fuel-air mixture prior to ignition. Fuel injection and inlet air swirl are coordinated to provide a small portion of rich mixture near the point of ignition surrounded by a large region of increasingly fuel-lean mixture. Flame propagation proceeds outward from the point of ignition through the leaner portions of the combustion chamber.

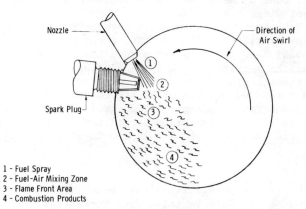

1 - Fuel Spray
2 - Fuel-Air Mixing Zone
3 - Flame Front Area
4 - Combustion Products

Figure 3.   Texaco-controlled combustion system (TCCS)

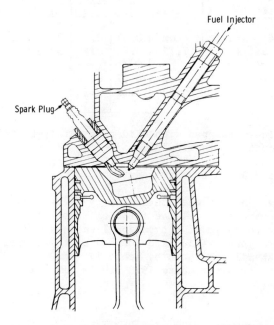

Figure 4.   Ford-programmed combustion (PROCO)
system

Both the TCCS and PROCO engines are inherently low emitters
of CO, primarily a result of lean mixture combustion. Unburned
HC and NO_x emissions have been found to be lower than those
typical of uncontrolled conventional engines, but it appears that
additional control measures are required to meet statutory 1976
Federal emissions standards.

The U.S. Army Tank Automotive Command has sponsored development of low emissions TCCS and PROCO power plants for light-duty
Military vehicles. These power plants have been based on conversion of the 4-cylinder, 70-hp L-141 Jeep engine. The vehicles
in which these engines were placed were equipped with oxidizing
catalysts for added control of HC and CO emissions, and EGR was
used as an additional measure for control of NO_x.

Results of emissions tests on Military Jeep vehicles equipped
with TCCS and PROCO engines are listed in Table II (10). At low

Table II

Average Emissions from Military
Jeep Vehicles with Stratified
Engine Conversions (Reference 10)

Engine	Miles	Emissions, g/Mile[2]			CVS Fuel Economy, mpg
		HC	CO	NO_x	
L-141 Ford[1]	Low	0.37	0.93	0.33	18.5-23
PROCO	17,123	0.64	0.46	0.38	
L-141 Texaco[1]	Low	0.37	0.23	0.31	16-22
TCCS	10,000	0.77	1.90	0.38	

[1]Engines equipped with oxidation catalysts and
exhaust gas recirculation.

[2]1975 CVS C-H test procedure.

mileage these vehicles met the statutory 1976 emissions standards.
Deterioration problems related to HC emission would be expected
to be similar to those of conventional engines equipped with
oxidizing catalysts. This is evidenced by the increase in HC
emissions with mileage shown by Table II. NO_x and CO emissions
appear to have remained below 1976 levels with mileage
accumulation.

Table III

Average Low Mileage Emissions
Levels - Ford PROCO Conversions
(Reference 10)

	Emissions,[2] g/Mile			CVS Fuel Economy, mpg	Inertia Weight, Lb
	HC	CO	NO_x		
PROCO 141-CID[1]	0.12	0.46	0.32	20.4	2500
Capri Vehicles	0.13	0.18	0.33	25.1	
	0.11	0.27	0.32	22.3	
PROCO 351-CID[1] Torino Vehicle	0.30	0.37	0.37	14.4	4500
PROCO 351-CID[1] Montego Vehicles	0.36	0.13	0.63	-	-
	0.36	1.08	0.39	12.8	-

[1]All vehicles employed noble metal exhaust oxidation catalysts and exhaust gas recirculation.

[2]1975 CVS C-H test procedure.

Table III presents emissions data at low mileage for several passenger car vehicles equipped with PROCO engine conversions (10). These installations included noble metal catalysts and EGR for added control of HC and NO_x emissions, respectively. All vehicles met the statutory 1976 standards at low mileage. Fuel consumption data, as shown in Table III, appear favorable when contrasted with the fuel economy for current production vehicles of similar weight.

Fuel requirements for the TCCS and PROCO engines differ substantially. The TCCS concept was initially developed for multi-fuel capability; as a consequence, this engine does not have a significant octane requirement and is flexible with regard to fuel requirements. In the PROCO engine combustion chamber, an end gas region does exist prior to completion of combustion; and, as a consequence, this engine has a finite octane requirement.

Prechamber Stratified
Charge Engines

A number of designs achieve charge stratification through division of the combustion region into two adjacent chambers. The emissions reduction potential for two types of dual-chamber engines has been demonstrated. First, in a design traditionally called the "prechamber engine," a small auxiliary or ignition

chamber equipped with a spark plug communicates with the much
larger main combustion chamber located in the space above the
piston (Figure 5). The prechamber typically contains 5-15% of
the total combustion volume. In operation of this type of
engine, the prechamber is supplied with a small quantity of fuel-
rich ignitable fuel-air mixture while a very lean and normally
unignitable mixture is supplied to the main chamber above the piston. Expansion of high temperature flame products from the prechamber leads to ignition and burning of the lean main chamber
fuel-air charge.

The prechamber stratified charge engine has existed in
various forms for many years. Early work by Ricardo (11) indicated that the engine could perform very efficiently within a
limited range of carefully controlled operating conditions. Both
fuel-injected and carbureted prechamber engines have been built.
A fuel-injected design initially conceived by Brodersen (12) was
the subject of extensive study at the University of Rochester for
nearly a decade (13,14). Unfortunately, the University of
Rochester work was undertaken prior to widespread recognition of
the automobile emissions problem; and, as a consequence, emissions characteristics of the Brodersen engine were not determined.
Another prechamber engine receiving attention in the early 1960's
is that conceived by R. M. Heintz (15). The objectives of this
design were reduced HC emissions, increased fuel economy, and
more flexible fuel requirements.

Initial experiments with a prechamber engine design called
"the torch ignition engine" were reported in the U.S.S.R. by
Nilov (16) and later by Kerimov and Mekhtier (17). This desig-
nation refers to the torchlike jet of hot combustion gases
issuing from the precombustion chamber upon ignition. In the
Russian designs, the orifice between prechamber and main chamber
is sized to produce a high velocity jet of combustion gases. In
a recent publication (18), Varshaoski et al. have presented emissions data obtained with a torch engine system. These data show
significant pollutant reductions relative to conventional engines;
however, their interpretation in terms of requirements based on
the Federal emissions test procedure is not clear.

A carbureted three-valve prechamber engine, the Honda
Compound Vortex-Controlled Combustion (CVCC) system, has received
considerable recent publicity as a potential low emissions power
plant (19). This system is illustrated schematically in Figure 6.
Honda's current design employs a conventional engine block and
piston assembly. Only the cylinder head and fuel inlet system
differ from current automotive practice. Each cylinder is
equipped with a small precombustion chamber communicating by
means of an orifice with the main combustion chamber situated
above the piston. A small inlet valve is located in each prechamber. Larger inlet and exhaust valves typical of conventional
automotive practice are located in the main combustion chamber.
Proper proportioning of fuel-air mixture between prechamber and

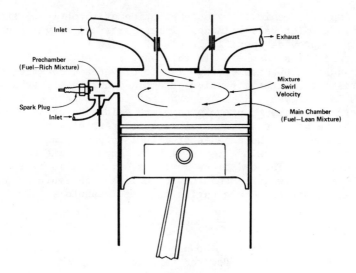

Figure 5. *Schematic of a prechamber charge engine*

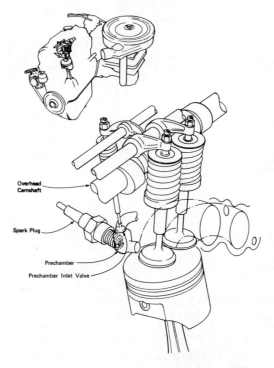

Environmental
Protection Agency

Figure 6. *Honda CVVC engine* (19)

main chamber is achieved by a combination of throttle control and
appropriate inlet valve timing. Inlet ports and valves are
oriented to provide specific levels of air swirl and turbulence
in the combustion chamber. In this way, a relatively slow and
uniform burning process giving rise to elevated combustion temperatures late in the expansion stroke and during the exhaust
process is achieved. High temperatures in this part of the engine
cycle are necessary to promote complete oxidation of HC and CO.
It should be noted that these elevated temperatures are neces-
sarily obtained at the expense of a fuel economy penalty.

Results of emissions tests with the Honda engine have been
very promising. The emissions levels shown in Table IV for
a number of lightweight Honda Civic vehicles are typical and
demonstrate that the Honda engine can meet statutory 1975-1976

Table IV

Honda CVCC
Powered Vehicle[1]
Emissions (Reference 19)

	Emissions,[2] g/Mile			Fuel Economy, mpg	
	HC	CO	NO$_x$	1975 FTP	1972 FTP
Low Mileage Car[3] No. 3652	0.18	2.12	0.89	22.1	21.0
50,000-Mile Car[4] No. 2034	0.24	1.75	0.65	21.3	19.8

[1]Honda Civic vehicles.

[2]1975 CVS C-H procedure with 2000-lb inertia weight.

[3]Average of five tests.

[4]Average of four tests.

HC and CO standards and can approach the statutory 1976 NO$_x$
standard (10). Of particular importance, durability of this
system appears excellent as evidenced by the high mileage emissions levels reported in Table IV. The noted deterioration of
emissions after 30,000-50,000 miles of engine operation was
slight and apparently insignificant.

Recently, the EPA has tested a larger vehicle converted to
the Honda system (20). This vehicle, a 1973 Chevrolet Impala
with a 350-CID V-8 engine, was equipped with cylinder heads and

induction system of Honda manufacture. Test results are presented
in Table V for low vehicle mileage. The vehicle met the present
1976 interim Federal emissions standards though NO_x levels were
substantially higher than for the much lighter weight Honda
Civic vehicles.

Table V

Emissions from Honda CVCC
Conversion of 350-CID
Chevrolet Impala (Reference 20)

Test	Emissions,[1] g/Mile			Fuel Economy, mpg
	HC	CO	NO_x	
1	0.27	2.88	1.72	10.5
2[2]	0.23	5.01	1.95	11.2
3[3]	0.80	2.64	1.51	10.8
4	0.32	2.79	1.68	10.2

[1]1975 CVS C-H procedure, 5000-lb inertia
weight.

[2]Carburetor float valve malfunctioning.

[3]Engine stalled on hot start cycle.

Fuel economy data indicate that efficiency of the Honda
engine, when operated at low emissions levels, is somewhat poorer
than that typical of well-designed conventional engines operated
without emissions controls. However, EPA data for the Chevrolet
Impala conversion show that efficiency of the CVCC engine meeting
1975-1976 interim standards was comparable to or slightly better
than that of 1973 production engines of similar size operating in
vehicles of comparable weight. It has been stated by automobile
manufacturers that use of exhaust oxidation catalysts beginning
in 1975 will result in improved fuel economy relative to 1973 production vehicles. In this event fuel economy of catalyst-
equipped conventional engines should be at least as good as that
of the CVCC system.

The apparent effect of vehicle size (more precisely the ratio
of vehicle weight to engine cubic inch displacement) on NO_x emissions from the Honda engine conversions demonstrates the generally
expected response of NO_x emissions to increased specific power
demand from this type of engine. For a given engine cubic inch
displacement, maximum power output can be achieved only by enrich-
ing the overall fuel-air mixture ratio to nearly stoichiometric

proportions and at the same time advancing ignition timing to the
MBT point. Both factors give rise to increased NO_x emissions.
This behavior is evidenced by Table VI, which presents steady
state emissions data for the Honda conversion of the Chevrolet
Impala (20). At light loads, NO_x emissions are below or roughly
comparable to emissions from a conventionally powered 1973 Impala.
This stock vehicle employs EGR to meet the 1973 NO_x standard. It
is noted in Table VI that for the heaviest load condition
reported, the 60-mph cruise, NO_x emissions from the Honda conversion approached twice the level of emissions from the stock
vehicles. This points to the fact that in sizing engines for a
specific vehicle application, the decreased air utilization (and
hence specific power output) of the prechamber engine when
operated under low emissions conditions must be taken into
consideration.

Table VI

Steady State Emissions from
Honda CVCC Conversion of
350-CID Chevrolet Impala (Reference 20)

Vehicle Speed, mph	Emissions, g/Mile					
	HC		CO		NO_x	
	350 CVCC	350 Stock	350 CVCC	350 Stock	350 CVCC	350 Stock
15	0.15	0.60	3.30	7.26	0.37	0.52
30	0.00	1.22	0.65	9.98	0.53	0.37
45	0.00	0.51	0.19	4.71	1.00	0.93
60	0.01	0.32	0.53	2.48	3.00	1.78

Divided-Chamber Staged
Combustion Engine

Dual-chamber engines of another type, often called "divided-
chamber" or "large-volume prechamber" engines, employ a two-stage
combustion process. Here initial rich mixture combustion and heat
release (first stage of combustion) are followed by rapid dilution
of combustion products with relatively low temperature air (second
stage of combustion). In terms of the concepts previously developed, this process is initiated in the vicinity of Point A of
Figure 2. Subsequent mixing of combustion products with air is
represented by a transition from Point A to Point B. The object
of this engine design is to effect the transition from Point A to
Point B with sufficient speed that time is not available for
formation of significant quantities of NO. During the second low
temperature stage of combustion (Point B), oxidation of HC and CO
goes to completion.

An experimental divided-chamber engine design that has been built and tested is represented schematically in Figure 7 (21,22). A dividing orifice (3) separates the primary combustion chamber (1) from the secondary combustion chamber (2), which includes the cylinder volume above the piston top. A fuel injector (4) supplies fuel to the primary chamber only. Injection timing is arranged such that fuel continuously mixes with air entering the primary chamber during the compression stroke. At the end of compression, as the piston nears its top center position, the primary chamber contains an ignitable fuel-air mixture while the secondary chamber adjacent to the piston top contains only air. Following ignition of the primary chamber mixture by a spark plug (6) located near the dividing orifice, high temperature rich mixture combustion products expand rapidly into and mix with the relatively cool air contained in the secondary chamber. The resulting dilution of combustion products with attendant temperature reduction rapidly suppresses formation of NO. At the same time, the presence of excess air in the secondary chamber tends to promote complete oxidation of HC and CO.

Results of limited research conducted both by university and industrial laboratories indicate that NO_x reductions of as much as 80-95% relative to conventional engines are possible with the divided-chamber staged combustion process. Typical experimentally determined NO_x emissions levels are presented in Figure 8 (23). Here NO_x emissions for two different divided-chamber configurations are compared with typical emissions levels for conventional uncontrolled automobile engines. The volume ratio, β, appearing as a parameter in Figure 8, represents the fraction of total combustion volume contained in the primary chamber. For β values approaching 0.5 or lower, NO_x emissions reach extremely low levels. However, maximum power output capability for a given engine size decreases with decreasing β values. Optimum primary chamber volume must ultimately represent a compromise between low emissions levels and desired maximum power output.

HC and particularly CO emissions from the divided-chamber engine are substantially lower than conventional engine levels. However, further detailed work with combustion chamber geometries and fuel injection systems will be necessary to fully evaluate the potential for reduction of these emissions. Table VII presents results of tests cited by the National Academy of Sciences (10).

Emissions from the divided-chamber engine are compared with those from a laboratory PROCO stratified charge engine, the comparison being made at equal levels of NO_x emissions. NO_x emissions were controlled to specific levels by addition of EGR to the PROCO engine and by adjustment of operating parameters for the divided-chamber engine. These data indicate that the divided-chamber engine is capable of achieving very low NO_x emissions with relatively low HC and CO emissions.

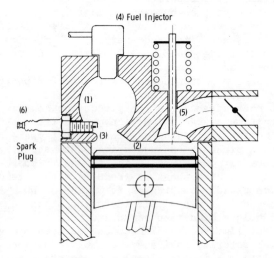

Figure 7. Schematic of a divided chamber engine (21)

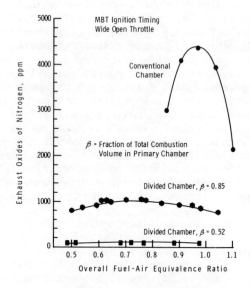

American Institute of Chemical Engineers

Figure 8. Comparison of conventional and divided combustion chamber NO$_x$ emissions (23)

Table VII

Single-Cylinder Divided Combustion
Chamber Engine Emissions
Tests (Reference 10)

Engine	NO_x Reduction Method	Emissions, g/ihp-hr			Fuel Economy, Lb/ihp-hr
		NO_x	HC	CO	
PROCO	EGR	1.0	3.0	13.0	0.377
Divided Chamber	None	1.0	0.4	2.5	0.378
PROCO	EGR	0.5	4.0	14.0	0.383
Divided Chamber	None	0.5	0.75	3.3	0.377

As shown by Table VII, fuel economy of the divided-chamber staged combustion engine is comparable to that of conventional piston engines without emissions controls. When compared with conventional piston engines controlled to equivalent low NO_x emissions levels, the divided-chamber engine is superior in terms of fuel economy.

The Diesel Engine

The diesel engine can be viewed as a highly developed form of stratified charge engine. Combustion is initiated by compression ignition of a small quantity of fuel-air mixture formed just after the beginning of fuel injection. Subsequently, injected fuel is burned in a heterogeneous diffusion flame. Overall fuel-air ratios in diesel engine operation are usually extremely fuel lean. However, major combustion reactions occur locally in combustion chamber regions containing fuel-air mixtures in the vicinity of stoichiometric proportions.

The conventional diesel engine is characterized by low levels of CO and light HC emissions, a result of lean mixture operation. On a unit power output basis, NO_x emissions from diesel engines are typically lower than those of uncontrolled gasoline engines, a combined result of diffusion combustion and, in an approximate sense, low mean combustion temperatures. Work devoted to mathematical simulation of diesel combustion has shown that NO formation occurs primarily in combustion products formed early in the combustion process, with the later portions of diffusion-controlled combustion contributing substantially less (24).

Table VIII presents emissions levels for three diesel-powered passenger cars as reported by the EPA (25). These vehicles, a Mercedes 220D, Opel Rekord 2100D, and Peugeot 504D, were powered by 4-cylinder engines ranging in size from 126-134 CID with power ratings ranging from 65-68 bhp. Two of the diesel-powered

vehicles were capable of meeting the 1975 statutory emissions
standards. NO_x emissions were in excess of the original Federal
1976 standards but were well within present interim standards.

Table VIII

Automotive Diesel Engine Emissions
(Reference 25)

Vehicle	Emissions, g/Mile				Inertia Weight, Lb	Fuel Economy, mpg	
	HC (Cold Bag)	HC (Hot FID)	CO	NO_x		1975 FTP	1972 FTP
Mercedes 220D	0.17	0.34	1.42	1.43	3500	23.6	23.3
Mercedes 220D (Modified)	0.13	0.28	1.08	1.48	3500	24.6	23.6
Opel Rekord 2100D	0.16	0.40	1.16	1.34	3000	23.8	23.2
Peugeot 504D	1.03	3.11	3.42	1.07	3000	25.2	24.2

The preceding data do not include information on particulate
and odorant emissions, both of which could be important problems
with widespread diesel engine use in automobiles. Complete
assessment of the environmental potential for the diesel engine
would have to include consideration of these factors as well as
emission of polynuclear aromatic hydrocarbons. All are the subject of ongoing research.

Fuel economy data referred to both 1972 and 1975 Federal test
procedures are presented in Table VIII. As expected, diesel
engine fuel economies are excellent when compared with gasoline
engine values. However, a more accurate appraisal would probably
require comparison at equal vehicle performance levels. Power-
to-weight ratios and, hence, acceleration times and top speeds
for the diesel vehicles cited above are inferior to values
expected in typical gasoline-powered vehicles.

Gas Turbine, Stirling Cycle,
and Rankine Cycle Engines

Gas turbine, Stirling cycle, and Rankine engines all employ
steady flow or continuous combustion processes operated with fuel-
lean overall mixture ratios. In a strict sense, the gas turbine
is an internal combustion engine since high temperature combustion
products serve as the cycle working fluid. Rankine and Stirling
engines are external combustion devices with heat exchanged
between high temperature combustion gases and the enclosed cycle
working fluid.

In contrast to the situation with conventional spark ignition piston engines, the major obstacles related to use of continuous combustion power plants are in the areas of manufacturing
costs, durability, vehicle performance, and fuel economy. The
problem of exhaust emissions, which involves primarily the combustion process, has been less significant than the foregoing
items.

As a consequence of lean combustion, these continuous combustion power plants are characterized by low HC and CO emissions.
Several investigators have reported data indicating that existing
combustion systems are capable of approaching or meeting statutory
1975 and 1976 vehicle emissions standards for HC and CO (26,27).

For a given power output, NO_x emissions appear to be lower
than those of conventional uncontrolled gasoline engines. However, it has been shown that existing combustors probably will
not meet the statutory 1976 NO_x standard when installed in motor
vehicles (26).

The formation of NO_x in continuous-flow combustors has been
found to result from the presence of high temperature zones with
local fuel-air ratios in the vicinity of stoichiometric conditions. Approaches suggested for minimizing NO_x formation have
involved reduction of these localized peak temperatures through
such techniques as radiation cooling, water injection, and primary zone air injection. Other approaches include lean mixture
primary zone combustion such that local maximum temperatures fall
below levels required for significant NO formation. Laboratory
gas turbine combustors employing several of these approaches have
demonstrated the potential for meeting the 1976 standards (28).
With a laboratory Stirling engine combustor, Philips has measured
simulated Federal vehicle test procedure emissions levels well
below 1976 statutory levels (29).

Conclusion

As an alternative to the conventional internal combustion
engine equipped with exhaust treatment devices, modified combustion engines can, in principle, yield large reductions in vehicle
exhaust emissions. Such modifications include stratified charge
engines of both open and dual chamber design. On an experimental
basis, prototype stratified charge engines have achieved low
exhaust emissions with fuel economy superior to that of conventional engines controlled to similar emissions levels.

The diesel engine is capable of achieving low levels of light
HC, CO, and NO_x emissions with excellent fuel economy. Potential
problems associated with widespread diesel use in light-duty
vehicles are initial cost, large engine size and weight for a
given power output, the possibility of excessive particulate and
odorant emissions, and excessive engine noise.

Several power plants based on continuous combustion processes
have the potential for very low exhaust emissions. These include

the gas turbine, the Rankine engine, and the Stirling engine. However, at the present time major problems in the areas of manufacturing costs, reliability, durability, vehicle performance, and fuel economy must be overcome. As a consequence, these systems must be viewed as relatively long range alternatives to the piston engine.

Literature Cited

1. Beckman, E. W., Fagley, W. S., and Sarto, J. O., Society of Automotive Engineers Transactions (1967), 75.

2. Cantwell, E. N., and Pahnke, A. J., Society of Automotive Engineers Transactions (1966), 74.

3. Bartholomew, E., Society of Automotive Engineers (1966), Paper 660109.

4. Campion, R. M., et al., Society of Automotive Engineers (1972), Publication AP-370.

5. Kopa, R. D., Society of Automotive Engineers (1966), Paper 660114.

6. Newhall, H. K., and Shahed, S. M., Thirteenth Symposium (International) on Combustion, p. 365, The Combustion Institute (1971).

7. Longwell, J. P., and Weiss, M. A., Ind. Eng. Chem. (1955), 47, p. 1634-1643.

8. Bishop, I. N., and Simko, A., Society of Automotive Engineers (1968), Paper 680041.

9. Mitchell, E., Cobb, J. M., and Frost, R. A., Society of Automotive Engineers (1968), Paper 680042.

10. "Automotive Spark Ignition Engine Emission Control Systems to Meet Requirements of the 1970 Clean Air Amendments," report of the Emission Control Systems Panel to the Committee on Motor Vehicle Emissions, National Academy of Sciences, May 1973.

11. Ricardo, H. R., SAE Journal (1922), 10, p. 305-336.

12. U.S. Patent No. 2,615,437 and No. 2,690,741, "Method of Operating Internal Combustion Engines," Neil O. Broderson, Rochester, New York.

13. Conta, L. D., and Pandeli, D., American Society of Mechanical Engineers (1959), Paper 59-SA-25.

14. Conta, L. D., and Pandeli, D., American Society of Mechanical Engineers (1960), Paper 60-WA-314.

15. U.S. Patent No. 2,884,913, "Internal Combustion Engine," R. M. Heintz.

16. Nilov, N. A., Automobilnaya Promyshlennost No. 8 (1958).

17. Kerimov, N. A., and Metehtiev, R. I., Automobilnoya Promyshlennost No. 1 (1967), p. 8-11.

18. Varshaoski, I. L., Konev, B. F., and Klatskin, V. B., Aubomobilnaya Promyshlennost No. 4 (1970).

19. "An Evaluation of Three Honda Compound Vortex-Controlled Combustion (CVCC) Powered Vehicles," Report 73-11 issued by Test and Evaluation Branch, Environmental Protection Agency, December 1972.

20. "An Evaluation of a 350-CID Compound Vortex-Controlled Combustion (CVCC) Powered Chevrolet Impala," Report 74-13 DWP issued by Test and Evaluation Branch, Environmental Protection Agency, October 1973.

21. Newhall, H. K., and El-Messiri, I. A., Combustion and Flame (1970), 14, p. 155-158.

22. Newhall, H. K., and El-Messiri, I. A., SAE Trans. (1970), 78, Paper 700491.

23. El-Messiri, I. A., and Newhall, H. K., Proc. Intersociety Energy Conversion Engineering Conference (1971), p. 63.

24. Shahed, S. M., and Chiu, W. S., Society of Automotive Engineers, Paper 730083, January 1973.

25. "Exhaust Emissions from Three Diesel-Powered Passenger Cars," Report 73-19 AW issued by Test and Evaluation Branch, Environmental Protection Agency, March 1973.

26. Wade, W. R., and Cornelius, W., General Motors Research Laboratories Symposium on Emissions from Continuous Combustion Systems, p. 375-457, Plenum Press, New York (1972).

27. Brogan, J. J., and Thur, E. M., Intersociety Energy Conversion Engineering Conference Proceedings, p. 806-824 (1972).

28. White, D. J., Roberts, P. B., and Compton, W. A.,
    Intersociety Energy Conversion Engineering Conference
    Proceedings, p. 845-851 (1972).

29. Postma, N. D., VanGiessel, R., and Reinink, F., Society of
    Automotive Engineers, Paper 730648 (1973).

Alternative Automotive Emission Control Systems

E. N. CANTWELL, E. S. JACOBS, and J. M. PIERRARD

E. I. du Pont Nemours and Co., Wilmington, Del. 19898

Abstract

Relationships developed between automotive exhaust emissions and ambient air quality levels have been used to establish the degree of emission control needed to meet the ambient air quality standards in major urban areas. Interim emission standards already established for 1975 appear to be more than adequate for congested urban areas and the existing 1974 standards appear more than adequate for the remainder of the country. The very low levels mandated by the 1970 Amendments to the Clean Air Act do not appear necessary. Emission control systems based on engine modifications and thermal reactors have been shown to provide the degree of emission control needed to achieve the ambient air quality goals and are compatible with the continued use of leaded gasoline. Exhaust lead traps have been shown to be a practical means to reduce lead emissions to the environment should such control be needed. The lead tolerant control systems used with high compression ratio engines are shown to have the potential to improve fuel economy to a greater degree than proposed catalytic systems used with unleaded gasoline and low compression ratio engines. Both approaches are shown to impose severe fuel economy penalties at the very stringent emission standards mandated by the 1970 Amendments to the Clean Air Act.

Introduction

The 1970 Amendments to the Clean Air Act directed the Environmental Protection Agency to set hydrocarbon and carbon monoxide exhaust emission standards for 1975 vehicles which were 90% lower than the standards for 1970 vehicles. These amendments also required a nitrogen oxides emission standard for 1976 vehicles which was 90% lower than the level of 1971 vehicles. In response to this mandate by Congress, EPA issued vehicle emission standards for 1975 and 1976 which were subsequently postponed one year to 1976 and 1977. At the time the standards were postponed, EPA issued interim emission standards for 1975 which called for reductions in hydrocarbons and carbon monoxide, but not to the degree required by the 1970 Amendments to the Clean Air Act. The 90% reduction required by the 1970 Amendments to the Clean Air Act coupled with the reductions achieved up to 1970 provided for overall reductions of 97% for hydrocarbons, 96% for carbon monoxide, and 93% for oxides of nitrogen below pre-emission control levels.

In April, 1971 EPA promulgated national ambient air quality standards for hydrocarbons, carbon monoxide, nitrogen dioxide, and photochemical oxidants as required by the 1970 Amendments to the Clean Air Act. These air quality standards were set to protect the public health and welfare from any known or anticipated adverse effects of air pollution.

Interestingly, automotive emission standards for hydrocarbons, carbon monoxide, and nitrogen oxides were set by Congress prior to establishment of ambient air quality standards. Actually, the process should have been reversed. Air quality standards should have been established first and then studies should have been carried out to determine what levels of automotive emissions were needed to meet the air quality standards. Since this was not done, we now find ourselves in the position of examining the legislated automotive standards to see if the degree of control being called for is consistent with the air quality goals. In the absence of any other consideration, it is desirable to attain the lowest automotive emission levels which are technologically achievable. Gasoline consumption and control hardware costs, however, increase with the degree of emission control imposed (1)*. In view of the critical energy situation, it is imperative that an appropriate balance be struck between required emission levels and fuel consumption penalties. Automotive emission standards must be soundly related to the ambient air quality standards, so that we are not paying for more control than is needed to achieve the established air quality standards.

The purpose of this paper is to report on the progress being made in major programs carried out at Du Pont designed to answer three questions derived from these concerns. These questions are:

- What automotive exhaust emission standards are needed to achieve the air quality standards?

- What automotive emission control systems can meet these needed automotive emission standards?

- Which emission control system minimizes consumption of energy while still meeting needed standards?

---

* Numbers in parentheses refer to references at end of paper.

Scope of Du Pont Studies

The process of selecting the proper course of action for air pollution control to achieve the objectives of the Clean Air Act consists of an orderly series of decisions as diagramed in Figure 1. First, pollutant effects on human health and welfare including the aesthetic qualities of life as well as damage to plants, animals, and inanimate objects of value must be determined. Using this information, Step A can be taken to establish the needed levels of air quality. Establishment of these air quality levels is not precise because all of the needed knowledge is not and probably never will be available. The degree of safety required is based on judgement and is subject to the interpretation of the adequacy and accuracy of the medical and environmental studies. Debates on these issues arise from the recognition that a larger factor of safety requires more stringent and thus more costly controls.

As discussed earlier, EPA set ambient air quality standards in 1971. The adequacy of these standards has been vigorously debated (2). Some critics contend that air quality standards are too high to protect public health while others contend that they are lower than can be supported based on medical evidence. Still others point out that some of the standards are below natural background levels and thus are impossible to attain. For the purposes of this paper, the national ambient air quality standards promulgated by EPA have been accepted as necessary goals for achieving clean air.

Once air quality standards have been set, then the question becomes one of deciding what degree of control of emission sources is needed to meet these standards. The ratio of existing and desired air quality levels gives the percentage reduction which is needed in source emissions. In order to carry out this part of the process, information is needed on air quality levels in critical areas of the country and a detailed knowledge of emission levels from various sources. At the time Congress enacted the Clean Air Act Amendments, information of this type was limited and Step B was side-stepped by Congress in enacting the degree of automotive controls required to meet ambient air quality standards. Since then, data have become available which make it possible to carry out this step and the first part of this paper presents a logical basis for establishing automotive emission standards.

The final step in the overall process is selecting the most effective means to meet emission standards. Once again this step is not precise because the costs in terms of economic, social, and environmental factors have not been quantified. This is due in part to inadequate data on alternative control measures. Identification of preferred control measures allows specific emission control strategies and technology to be scheduled for implementation.

The third and final section of this paper provides an assessment of the effect of various emission control systems on energy consumption using the most recent data on vehicle fuel economy. The uncertainties inherent in Steps A and B, coupled with the uncertainty in Step C prevent quantitative assessment of the cost effectiveness of various implementation plans. Because of this, it is vital that each alternative be examined thoroughly before a final decision on implementation is made.

The purpose of this paper is to provide an analysis of available data to arrive at a reasonable strategy for automotive emission control to achieve air quality to protect public health.

Air Quality and Emission Standards

In order to calculate automotive emission standards needed to meet air quality standards, Du Pont analyzed data amassed since 1962 under the EPA's Continuous Air Monitoring Program (CAMP) for gaseous pollutants in the central business districts of six large cities. Carbon monoxide was chosen for initial study because, unlike other motor vehicle exhaust pollutants, most carbon monoxide in urban areas comes from motor vehicles, and it is chemically inert in the atmosphere. In our studies, we have examined various proposed relationships between automotive emission rates and atmospheric levels of carbon monoxide. It was recognized that once a relationship was established for carbon monoxide, it then could be extended to other pollutants.

Simple Rollback Approach. Practical estimates of the automotive emissions reductions needed to achieve ambient air quality standards have been made on the basis of the so-called "rollback" approach first used by California and later by EPA. This approach assumes that the pollutant concentration is proportional to the emission rate of that pollutant in an air basin, with a small correction for the background level of the pollutant.

We tested this assumption for Chicago, using data from the EPA CAMP station in the central business district, and from outlying air sampling stations operated by the City of Chicago. Two of these sites were in residential areas, and the other two near expressways, at distances from 2.5 to 12 miles from the center city CAMP station. The analysis, described in detail in Reference 3, showed that daily average CO concentrations at the five stations were essentially unrelated. This indicated that ambient carbon monoxide levels should be related to nearby, not remote, traffic activity. Therefore, Chicago central business district traffic, rather than metropolitan area traffic, must be considered to explain air quality measured at the CAMP station.

This analysis also pointed up a problem in applying rollback, namely the proper choice of the traffic growth factor. Cordon traffic counts verified that traffic saturation exists in the Chicago central business district around the CAMP station, as illustrated in Figure 2. No large increase of vehicle use can reasonably be expected in other large urban business districts either. Therefore, it appears that simple rollback cannot be applied to develop projections of air quality in mature urban regions on the basis of anticipated growth in fringe regions.

Modified Rollback. Because of the limitations of simple rollback, recent EPA air quality projections have been based on a "modified" roll-back model, as discussed in Reference 4. The modifications introduced by EPA were:

1. consideration of six source categories, each with its own growth rate and assumed degree of future emission control, and

2. inclusion of a factor for each source category which weights the emissions according to their stack height.

With EPA's cooperation in providing us with copies of their input data and computer program, we have conducted tests of the ability of the modified rollback model to predict air quality. This was done by taking as a starting point, a measured air quality condition of, say, five years ago, and then using this value to calculate present air quality using the modified rollback model. The modified rollback model does not appear to be a good predictor since calculated values did not agree with measured values as shown in Figure 3.

Based on our earlier analysis of the relationship between urban CO emissions and air quality, we believe the inability of the modified rollback to predict is due to the assumption which calls for equal weighting of CO emissions originating anywhere within the air quality control region (AQCR). The Federal AQCR's are geographically very large, and therefore sources near the boundaries cannot be expected to have any significant influence on air quality in the central metropolitan area. For example, the areas of the Chicago, Philadelphia, and National Capitol AQCR's are, respectively, 6,087, 4,585, and 2,326 square miles. If these regions were circular, their outer boundaries would be 44, 38, and 27 miles from the center. As the regions are not circular, their farthest points are at even greater distances than those listed above. Yet, the simple rollback, and the current modified rollback assume that all sources contribute to air quality at a receptor point in proportion to their emission rate, regardless of distance from the receptor.

Air Quality Trend Analysis. There is an alternative to the simple rollback approach which can be used to establish needed vehicle emission

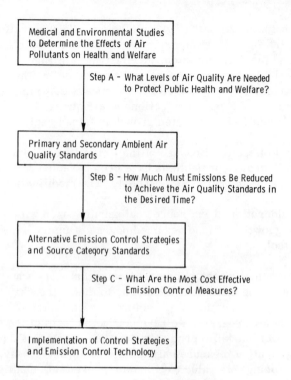

Figure 1. Sequence of steps required to achieve control of air pollution

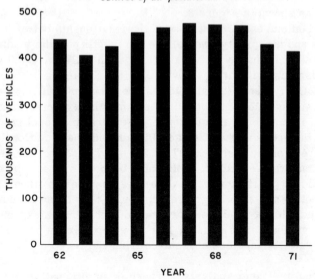

Figure 2. Weekday traffic activity in the Chicago central business district

standards to satisfy air quality standards. This approach, developed in our studies, and described in References 3 and 4, uses aerometric data records for cities with long term measurement bases, to establish what air quality changes have taken place over the years. By relating these changes to vehicle CO emission rates it is possible to predict needed vehicle emission rates to meet air quality standards.

An examination of CAMP data over the last ten years shows a general improving trend in CO air quality. Over the same period of time the average CO emission rate of vehicles on the road has decreased. This decrease has resulted from retirement of older, uncontrolled cars, and their replacement with new models, which have met increasingly stringent emission standards.

In the preceeding section, it was shown that Chicago central business district traffic is constant. This condition of traffic saturation is generally accepted as characteristic of mature center city districts, which coincide with the regions of expected maximum CO concentration, and with the locations of CAMP sampling stations. Because traffic can be considered constant, the value of average CO emission rate per mile can be used as an index of total CO emissions in the vicinity of the air monitoring station. A comparison of average vehicle CO emission rate and the highest 8-hour nonoverlapping average CO concentration in each month of record for the Chicago CAMP station is shown in Figure 4.

A projection of the trends shown in Figure 4 predicts that a value of 40 g/mile average CO emission rate for all vehicles on the road would allow attainment of the CO air quality standard for Chicago in the mid-1970's. While there is no reason to disagree with this finding, a comprehensive analysis of the predictive value of this approach to determination of emission standards, given in Reference 3, shows that projections on the basis of such trend envelopes are subject to large uncertainty. The sensitivity of trend line location to a few extreme values led us to the search for a more stable parameter describing air quality, which is discussed in the next section.

Use of Annual Mean for Predicting Emission Standards. Air quality standards for CO, hydrocarbons, and photochemical oxidant are specified as concentration levels, over given averaging times, not to be exceeded more than once a year. Therefore, demonstration that the standards are being achieved, or evaluation of progress toward achievement of the standards, requires a high degree of confidence that the air quality measurements used to estimate the frequency distribution of pollutant concentrations are representative. Unless truly continuous measurements are made, the confidence of an estimate of the frequency of occurrence of a rare event, such as is implied by the air quality standard, is degraded. However, if it could be shown that a definite relationship exists between

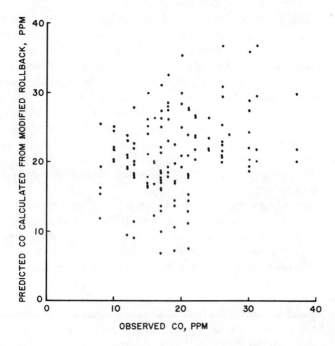

*Figure 3.   Comparison of carbon monoxide concentration pro-
jected by "modified rollback air quality impact analysis" with
observed concentrations. Annual second highest 8-hr. nonoverlapping average value.*

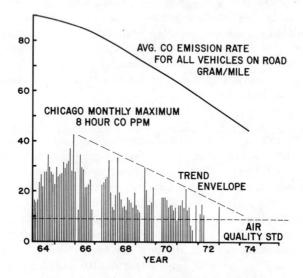

*Figure 4.   Comparison of vehicle carbon monoxide
emission rate and monthly maximum carbon monoxide
values in Chicago*

the number of violations of the concentration level of the standard, and a less sensitive property of the distribution, such as a long period mean, then that mean can be used to determine appropriate emission reductions needed to meet air quality standards.

By analysis of an aggregate of 51 years of data from the six CAMP cities, we established that a relationship does hold between the number of times per year that the air quality standard is exceeded and the annual average CO concentration.   The details of the analysis are reported in Reference 3.   From this relationship, the annual average CO which will result in only one annual occurrence of the 9 ppm - 8-hour average CO concentration was computed.   This value is 2.3 ppm.   An independent, strictly empirical analysis also gives a value of 2.3 ppm, with better than 98% confidence.

The annual average CO data listed in Table 1 show that Cincinnati and Philadelphia were close to, or already at, the required 2.3 ppm level in 1971.   The other cities were still above the required 2.3 ppm annual average by varying amounts, although trending downward.

The only exception, Denver, has shown an increase of ambient CO since 1969, which apparently is attributable to an excessive vehicle population mean CO emission rate resulting from the interpretation of the antitampering provision of the Clean Air Act Amendments to mean that altitude-tuning of carburetors is prohibited.   Vehicles at Denver's 5,000 foot elevation are running richer resulting in higher CO emissions. This explanation is verified by a recently reported EPA-sponsored survey of exhaust emissions from a total of 1,020 vehicles of the 1957-1971 model years which showed that hydrocarbon and carbon monoxide emissions in Denver were significantly higher, and oxides of nitrogen emissions significantly lower, than in the other cities studied (5).   As shown in Table 2, the Denver 1971 vehicle population mean CO emission rate is nearly double that for Washington, D. C.

The second column in Table 2 was calculated from the reported 1971 annual mean CO for each city, and the requirement for a 2.3 ppm annual average to meet the air quality standard as derived earlier.   When the percent reduction figures from column two are applied to the population mean CO emission rates in column one, the values in the last column are obtained, showing the CO emission rate consistent with achievement of the ambient air quality standard in each of the cities.   There is good agreement among the values for Denver, Los Angeles, St. Louis, and Washington.   The more stringent standard calculated to be necessary to meet the air quality standard in Chicago may well be the result of the meteorological peculiarity of that city's pronounced lake breeze effect discussed in Reference 3.

For the worst case city, Chicago, a vehicle CO emission rate of 29 g/mile is needed to meet air quality standard.   This value, obtained

Table 1

CAMP City Ambient Carbon Monoxide Trends

Year	Annual Average CO Concentration, ppm					
	Chicago	Denver	Washington	Cincinnati	St. Louis	Philadelphia
1962			5.3			
1963			6.9	7.1		
1964	12.1		5.7	6.1	6.4	7.2
1965	17.1	7.3	3.7	4.0	6.5	8.1
1966	12.5	7.9	3.3	4.9	5.8	6.8
1967	8.8	7.6	4.9	5.6	5.6	6.4
1968	6.2	5.4	3.4	5.7	4.6	8.7
1969	8.2	4.6	3.0	-	5.1	3.5
1970	6.9	6.5	3.8	-	4.4	4.1
1971	5.4	6.7	3.5	2.3	4.4	2.6
1972	-	6.3	-	-	4.4	-

Table 2

Vehicle CO Emission Standards Calculated By Air Quality Trend Analysis, Using Measured 1971 Vehicle Population Mean CO Emission Rates

City	CO Emission Rate Of Cars On Road In 1971, g/Mile	Reduction of 1971 Annual Mean Ambient CO Required to Meet Air Quality Standard, %	Vehicle CO Emission Rate Needed To Meet Air Quality Standard, g/Mile
Chicago	66	57	29
Denver	112	66	38
Los Angeles	74	52	36
St. Louis	75	48	39
Washington, D.C.	62	34	41

by statistical treatment of all the air quality data, is a more reliable value than that derived from Figure 4, where the trend line is dependent on only a few peak values.

Traffic Volume and Allowable CO Emission Rate. Another approach which can be taken to establish vehicle emission standards consists of treating an urban area as a box into which is injected carbon monoxide. The amount injected is directly proportional to the number of vehicle miles travelled in the area and the vehicle emission rate. This amount, defined as the CO flux density, is responsible for the ambient level of CO which occurs. The ratio of the observed CO level and the desired 9 ppm ambient air standard determines the degree of control or reduction in flux density which is needed. This would then apply for existing traffic conditions. If traffic saturation is then assumed, a vehicle standard can be derived which would apply in the worst case. This approach has been applied using data from a number of metropolitan areas and a value of 26 grams of CO per mile to meet the ambient air quality standard was obtained. Details of this development are covered in the following paragraphs.

Traffic volume data for the most heavily trafficked sections of each metropolitan area listed in Table 3, except Chicago, were obtained from EPA Transportation Control Strategy reports (6). For six of these areas, the authors of the individual reports had computed values of CO emission flux density. We calculated CO emission flux densities for Chicago, Denver, and Philadelphia using reported area traffic volume data (3) (6), appropriate values for the 1971 population mean CO emission rate (5), and the equation

$$F = 10^{-6} \, VR$$

where F = CO Emission Flux Density in

$$\frac{\text{metric tons of CO}}{\text{square miles x day}}$$

V = Area Traffic Volume in

$$\frac{\text{vehicle miles}}{\text{square miles x day}}$$

R = Mean CO Emission Rate in

$$\frac{\text{grams of CO}}{\text{vehicle mile}}$$

The air quality used in this analysis was the annual second-highest 8-hour running mean CO value (computed hourly) for each location, obtained from the Transportation Control Strategy reports and the CAMP data described in Reference 3. Assuming a linear relationship between CO emission flux density and ambient CO, we calculated the needed CO emission flux density to achieve the 9 ppm CO air quality standard for each area, as listed in the last column of Table 3.

Table 3

CO Emission Flux Density Needed For
Achievement Of Running Mean
Eight-Hour CO Air Quality Standard

City	Year	F Actual	Ambient 8-Hour CO, ppm,	F Needed
Boston				
Kenmore Square	1970	13.18	16.0	7.41
Science Park	1970	14.19	18.4	6.94
Chicago	1971	19.91	21.0	8.53
Denver	1971	20.93	27.5	6.85
Minneapolis	1971	17.09	17.5	8.79
Philadelphia	1970	16.86	19.0	7.98
Pittsburgh	1972	23.26	21.2	9.88
St. Paul	1971	16.83	21.6	7.01
Spokane	1970-71	15.71	20.0	7.07
			Average	7.83
			S. D.	1.05

The spread of values is small, considering the variety of locations and the fact that the air quality values range from less than two to more than three times the 9 ppm level. This indicated that the mean value of CO emission flux density could be used as the basis for deriving the vehicle population mean CO emission rate needed to achieve the air quality standard as a function of traffic volume.

Given the value of CO emission flux density, F, of 7.83, the vehicle population mean CO emission rate, R (grams/mile), needed to meet the air quality standard was calculated for various levels of daily average area traffic volume, V (vehicle miles traveled per square mile per day). The results of these calculations are given in Table 4 for four values of daily average area traffic volume ranging up to the estimated traffic saturation value of 300,000 vehicle miles per day per square mile (3).

These vehicle emission rates would meet the 8-hour CO air quality standard on the running mean basis. At traffic volumes below the saturation value, higher vehicle population mean CO emission rates would still permit achievement of the 8-hour running mean air quality standard. The area traffic volume in the most heavily traveled portions of the most metropolitan centers is below 200,000 vehicle miles per square mile per day as shown in Table 5. These extreme levels of traffic activity are encountered in only a few congested urban areas of the country.

Table 4

Needed Vehicle CO Emission Rate
To Meet Air Quality Standard As A
Function Of Traffic Volume

Traffic Volume, Vehicle Miles Per Day Per Square Mile	Vehicle CO Emission Rate Needed To Meet Air Quality Standard, g/Mile
150,000	52
200,000	39
250,000	31
300,000	26

Table 5

Daily Area Traffic Volumes In Heavily
Trafficked Regions Of Metropolitan Centers

City	Year	Area Traffic Volume, Vehicle Miles Per Day Per Square Mile
Boston		
Kenmore Square	1970	159,000
Science Park	1970	172,000
Chicago CBD	1971	300,000
Denver CBD	1971	187,000
Minneapolis CBD	1971	207,000
Philadelphia CBD	1970	234,000
Pittsburgh CBD	1972	281,000
St. Paul CBD	1971	204,000
Spokane CBD	1970-71	190,000

Calculations then were made to predict how the value of the CO emission standard chosen for 1976 and later will affect the year in which the air quality standard for CO would be met in the most heavily trafficked portions of urban areas. First, the future vehicle population mean CO emission rates were computed for each year through 1985 for a series of levels of the 1976 automotive emission standard for CO according to the procedure described by EPA (7). Values from this reference were used for projected vehicle replacement rates, annual miles traveled, emission rate, and deterioration of emission control by vehicle model year and age.

By this procedure a family of curves - one for each assumed 1976 CO
emission standard - was developed showing the trend of vehicle population mean CO emission rate as a function of calendar year. From
this family of curves, it was determined when the mean CO emission
rate would decrease to the needed level for a given value of area traffic
volume, assuming introduction of various CO emission standards beginning with the 1976 models. These values of CO air quality standard
attainment date versus 1976 CO emission standard are plotted in Figure
5 for four assumed levels of traffic activity ranging up to 300,000 vehicle
miles per day per square mile.

As Figure 5 illustrates, the attainment date of the CO air quality
standard is only slightly advanced by a 1976 CO emission standard lower
than the current standard of 28 grams per mile for areas with average
daily area traffic volumes up to 200,000 vehicle miles per square mile
per day. At higher levels of traffic volume, the attainment date does
advance with lower 1976 standards, but only at the estimated saturation
level is there a marked improvement in attainment date with a 1976 CO
emission standard as low as 15 grams per mile.

It is important to recognize that the extremes of traffic volumes are
encountered only in a limited number of locations. In general, only the
most heavily-trafficked square mile or two of a metropolitan area has
these area traffic volume values. Another point to be kept in mind in
interpreting Figure 5 is that, even with a 1976 emission standard of 3.4
g/mile, the air quality standard will not be achieved before 1978 in areas
of near-saturation traffic activity. Therefore, only through traffic control to reduce traffic volume to roughly 200,000 vehicle miles per square
mile per day can the air quality standard be achieved by the statutory
deadline of 1977. Once such traffic control measures are in force, no
advantage is gained from a 1976 emission standard lower than the current
level of 28 g/mile.

Implication of Air Quality Studies. The air quality trend analysis for
CO indicated a level of 29 g/mile would be satisfactory to meet ambient
CO air quality standard. A second method based on traffic saturation
conditions and maximum CO flux density in major urban areas resulted
in a value of 26 g/mile. In view of the large difference between these
values and the 3.4 g/mile mandated by the Amendments to the Clean Air
Act of 1970, a serious review of the standards is in order

Although our results to date have been restricted to the case of CO,
others have reexamined the HC and NO_x emission standards in relation
to the air quality standards for NO_2 and photochemical oxidants. In the
joint report by the panels on emission standards and atmospheric chemistry of the Committee on Motor Vehicle Emissions of the National Academy
of Sciences (8), it is concluded that the Federal emission standards of

0.41 gram/mile for HC and 0.4 gram/mile for NO_x seem too stringent by a factor of about 3. The panels estimated that emission rates needed to achieve the NO_2 and oxidant air quality standards should be of the order of 1.3 grams/mile of HC and 1.5 grams/mile of NO_x.

However, in an appendix to the joint panel's report, the argument is developed that very substantial reductions of hydrocarbon emissions may be required to achieve the oxidant air quality standard. This is because laboratory studies show that NO acts as an inhibitor, and NO_2 as a promoter of the photochemical smog reaction, and NO is converted to NO_2 when hydrocarbons and sunlight are present. Therefore, since most NO_x enters the atmosphere as NO rather than as NO_2, a conceivable strategy for oxidant standard achievement is hydrocarbon emission reduction without NO emission reduction.

In either event, the mandated 0.4 gram/mile NO_x emission standard appears more stringent than necessary. Further evidence of this excessive stringency is furnished by the EPA statement that because faulty analytical techniques led to incorrect ambient NO_2 levels, the vehicle NO_x emission standard should be revised upward (9). Revision of this standard would significantly widen the scope of choices among options for future vehicle emission control. Among the expected benefits of revised standards – still consistent with achievement of the air quality standards – would be improved system reliability, reduced maintenance costs, and improved fuel economy. The later sections of this paper deal with these considerations.

Alternate Automotive Emission Control Systems

Automobile exhaust emission standards required by the 1970 Amendments to the Clean Air Act as well as the 1975 interim standards are lower than the emission levels from current cars. The domestic automotive manufacturers have indicated they will use catalytic devices to achieve these interim as well as the statutory automotive emission standards. These catalytic systems, however, will require lead-free gasoline. Anticipating the use of catalysts and the need for lead-free gasoline, the engine compression ratios of cars built since 1970 were reduced to allow use of lower octane unleaded gasoline without knock. This compression ratio reduction has resulted in higher vehicle fuel consumption. Thus, the decision to use catalysts requiring unleaded fuel resulted in increased use of refinery crude oil because of a decrease in vehicle fuel economy and the need to make unleaded gasoline.

On the other hand, if the standards recommended in the previous section were adopted, then alternative emission control systems could be considered. As an example, the emission control systems based on thermal reactors can be used. These non-catalytic systems are compatible with leaded fuel and high compression ratio engines and could be used to give improved fuel economy. Thus, it is desirable to consider the performance of alternate control systems capable of meeting the less stringent standards, particularly in view of the fuel savings which may be possible.

Du Pont Total Emission Control System. The Du Pont emission control system consists of three major elements and appropriate engine modifications to optimize emission control and fuel economy. As indicated in Figure 6, the first component consists of an exhaust manifold thermal reactor which replaces the conventional exhaust manifold. The reactor provides a high temperature zone where exhaust gases mix with air supplied by an air pump and in which the hydrocarbons and carbon monoxide are oxidized to carbon dioxide and water. The second component consists of an exhaust gas recirculation (EGR) device which, when coupled with modifications to the carburetor metering and ignition timing, controls nitrogen oxides. The EGR device returns a portion of the exhaust gas to the carburetor to dilute the air-fuel mixture, which in turn reduces peak combustion temperatures and reduces nitrogen oxides formation. Finally, a muffler lead trap is used in place of the conventional exhaust muffler to remove most of the particulate lead from the exhaust gas before it leaves the tail pipe. When combined, these components form the Du Pont Total Emission Control System or TECS.

First and second generation thermal emission control systems, TECS I and TECS II, were described in earlier publications (10) (11). The TECS I vehicles were designed in 1969 and 1970 to meet the then existent exhaust emission standards for the 1975 model cars set forth by the U.S. government and by the State of California. After the passage of the 1970 Amendments to the Clean Air Act, a second generation system, TECS II, was developed in an attempt to meet the more stringent emission levels mandated by the Act. More recently, the TECS III and IV systems have been developed to meet emission levels compatible with attainment of the ambient air quality standards while minimizing the fuel consumption penalty.

The first generation emission control system, TECS I, was successful in meeting the goals of the earlier 1975 U.S. and California standards. Six 1970 Chevrolets equipped with TECS I were operated in a 10-month field test by the California Air Resources Board. The fleet average emission levels remained below the former 1975 standards throughout this test. Extended operation of the TECS I vehicles revealed two major problems. First, fuel consumption was significantly higher than comparable unmodified production 1970 vehicles. During the 10-month test conducted in California, TECS I vehicles used 17% more fuel than unmodified production vehicles. The second problem was excessive wear

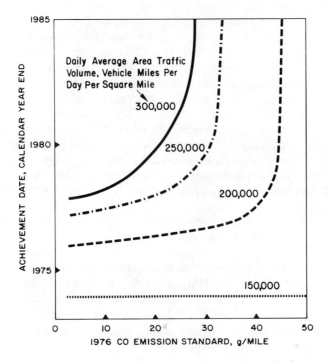

Figure 5. Date of achievement of air quality standard for carbon monoxide as a function of vehicle emission standards and traffic activity

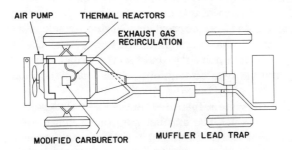

Figure 6. Du Pont total emission control system (TECS)

on various parts of the engine caused by the recirculation of metal oxides from the exhaust system back into the engine with the exhaust gas recirculation gases. This excessive wear problem was overcome by system modifications employed in TECS III and IV.

Vehicles equipped with the second generation emission control system, TECS II, included a full-size Chevrolet sedan equipped with a V-8 engine and automatic transmission, and a sub-compact Pinto equipped with a small four-cylinder engine and manual transmission. These vehicles approached but did not meet the emission levels required by the 1970 Amendments to the Clean Air Act. Hydrocarbon levels were below 0.25 gram per mile but carbon monoxide levels were about 6 grams per mile and nitrogen oxide levels were about 0.6 gram per mile. To meet the very low nitrogen oxide levels, carbutetors were operated very rich which increased fuel consumption. The modified Chevrolet used 35% more fuel and the Pinto 20% more fuel when compared with respective unmodified production vehicles.

In view of the growing concern over the increased consumption of fuel by low emission cars, it was decided to develop improved total emission control systems and evaluate internal engine modification for emission control. The goal of this program was to achieve emission levels adequate to meet ambient air quality standards in major urban areas while minimizing the fuel consumption penalty which usually accompanies the attainment of such low emission levels. The designs, emission control, and fuel economy achieved with these systems are discussed in succeeding sections.

Design of TECS III. The same general emission control systems used for TECS I and II were used in TECS III and IV. The major difference in the later systems is the emphasis on fuel economy and the use of high compression ratio engines with TECS IV vehicles.

A 1971 Pinto and a 1971 Chevrolet were chosen for the development of the TECS III because of our experience with these vehicles and the fact that they represented both current production standard size vehicles with V-8 engines and automatic transmissions as well as the sub-compact cars representative of both domestic and foreign production. A description of the vehicles and the type of control system installed is summarized in Table 6.

The TECS III system installed on the 1971 Chevrolet is shown schematically in Figure 6. The exhaust manifold reactors shown in Figure 7 were modifications of the Type VIII insulated reactor previously described (11). The interior core of the reactor was a large diameter open chamber with a single baffle located across the outlet pipe. The exhaust gases entered the core through extended exhaust ports and exited from these ports through rectangular slots which contain directional vanes.

The core consisted of concentric, double walled cylinders to reduce the heat loss from the inner core to the outer shell of the reactor. To further reduce heat loss, the outer shell and outlet pipe from the reactor were covered with a one-half-inch layer of fibrous ceramic insulation. Secondary air was injected into the exhaust ports and into the ends of the reactor to provide an overall oxidizing atmosphere.

Table 6

TECS III Vehicles

	1971 Chevrolet	1971 Pinto
Vehicle	Four-door sedan, 4,500 lbs	Two-door hardtop, 2,250 lbs
Engine	350 CID V-8	1600 cc In-line 4 cylinders
Transmission	3 Speed automatic	4 Speed manual
Thermal Reactor	Modified Type VIII, insulated, with air injection	Type V, shielded with air injection
EGR	Below throttle, ~10%	Above throttle, 13%
Traps	None	Muffler lead trap
Carburetion	Moderately enriched, fast release choke	Moderately enriched, fast release choke
Spark Timing	Modified vacuum advance on start up	Modified vacuum advance on start up

The exhaust gas recirculation system for the Chevrolet TECS III vehicle utilized two 1973 production EGR valves for this make vehicle. Exhaust gas was taken from the exhaust pipe ahead of the muffler on one side of the dual exhaust system, routed through the two EGR valves in parallel, and introduced below the secondary throttle plates of the four-barrel carburetor. The EGR valves were operated by manifold vacuum obtained from a port in the carburetor and were turned off at idle and at wide-open-throttle.

Changes were made to the carburetor metering, spark advance characteristics, and exhaust gas recirculation rates of the 1971 Chevrolet to improve fuel economy while maintaining low exhaust emission levels. A modified version of the production four-barrel carburetor normally used for this engine was employed. Metering was set to give a maximum vacuum air-fuel ratio at idle and a gradual leaning of the air-fuel ratio at mid-speed cruise range to give air-fuel ratios of 13.5 to 14:1. A production 1970 distributor was used with the basic timing advanced 8° from the manufacturer's specifications for the 1970 model to improve fuel economy. The production intake manifold was replaced with a dual-passage Offenhauser manifold and the intake system heat was supplied by engine coolant routed directly from the outlet of the engine through the internal passages in the intake manifold.

The TECS III system installed on the 1971 Pinto is shown schematically in Figure 8. The thermal reactor, shown in Figure 9, was a Type V exhaust manifold reactor. Appropriate modifications to the reactor port spacing, flanges, and the exhaust outlet were made to fit this design to the 1.6 liter Pinto engine. The reactor contained an inner core to promote mixing of the air and the exhaust gases and a double walled heat shield to reduce heat loss to the outer shell. No external insulation was used. All interior parts were made of Inconel 601 and the outer shell was 304 stainless steel.

The air injection system utilized a single air pump driven by a belt off the crankshaft at a pump-to-engine speed ratio of 0.75:1 (the output varied from about 3 cfm at idle to 15 cfm at 55 mph). The air from the pump was injected into each exhaust port.

The exhaust gas recirculation system for the Pinto is shown schematically in Figures 9 and 10. The exhaust gas was withdrawn from the exhaust ports in the immediate vicinity of the exhaust valve. The individual lines from each of the four ports were manifolded together to take advantage of the exhaust blowdown process to force the exhaust gas through the EGR line. A vacuum operated on-off valve prevented exhaust gas recirculation during cold start up. Exhaust gas was delivered to the carburetor below the venturi and above the throttle plate. The air injection tubes were located approximately one inch downstream from the exhaust gas recirculation pick up tubes in the exhaust ports. The recirculated exhaust contained about 25% injected air. A total of 18% exhaust gas and injected air was recirculated to the carburetor resulting in a true exhaust gas recirculation rate of 13% in terms of the air-fuel mixture delivered to the engine.

The carburetor used on the Pinto was selectively enriched throughout the metering range to aid in control of nitrogen oxide emissions. The air-fuel ratio varied from 13.5 to 14.0:1 over mid-speed range. The idle speed was increased from a standard 900 to 1000 rpm. A vacuum spark advance control system sensitive to coolant temperature was incorporated to prevent manifold vacuum advance during cold starts until the coolant reached normal operating temperatures.

Design of TECS IV. The TECS IV system was installed on a 1973 Chevrolet equipped with a 350 CID, V-8 engine as shown in Figure 6. The engine compression ratio was increased to 9.25:1 by use of 1970 350 CID engine heads. Otherwise, TECS IV was quite similar to the TECS III installed on the low compression ratio 350 CID engine of the 1971

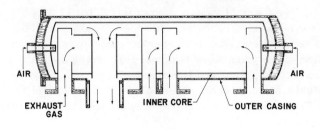

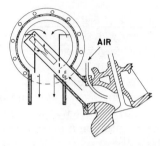

Figure 7. Modified Type VIII thermal reactor for 350 CID Chevrolet

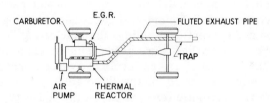

Figure 8. Du Pont TECS III installed on a 1.6 liter Pinto vehicle

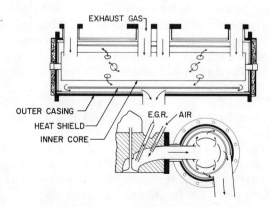

Figure 9. Type V thermal reactor for a 1.6 liter Pinto

Chevrolet. The thermal reactor shown in Figure 11 was a Type IX. The interior core was changed to enhance exhaust gas mixing to provide improved combustion under lean operating conditions. Again, a modified version of the production 4-barrel carburetor normally used with this engine was employed and the metering was set to give lean operation with air-fuel ratio of approximately 15:1. The basic ignition timing was advanced 10° from the manufacturer's specification to improve fuel economy.

Emission Control With TECS III and IV. The Pinto equipped with the Du Pont TECS III system was driven successfully for 100,000 miles using leaded gasoline. The road mileage accumulation consisted first of a 4,000-mile, cross-country trip from the east coast to Denver, to the continental divide, and back to the east coast. For the remainder of the test, the vehicle was driven during the daytime on an urban-suburban driving route in the South Jersey area. During the evening and nighttime hours, mileage was accumulated on a programmed chassis dynamometer following a simulated turnpike driving schedule with speeds varying between 40 and 60 mph with an average speed of 50 mph. An unmodified production version of the 1971 Pinto was run as a companion car with all of its mileage being accumulated in a similar manner except for the cross-country road trip. Commercial gasoline containing 2.2 grams of lead per gallon and conventional multigrade lubricating oils meeting the manufacturer's specifications were used for these tests.

The gaseous exhaust emission levels of hydrocarbons, carbon monoxide, and nitrogen oxides for the TECS III Pinto measured at intervals during the 100,000-mile test are shown in Figure 12. As can be seen, the average emission levels were below both the 1975 U.S. and the more stringent California interim emission standards for the entire test. The average emission levels during the 100,000-mile test as measured by the 1975 CVS Federal Test Procedure are given in Table 7.

Mechanical failures such as burned valves and worn clutches occurred in both the companion unmodified and TECS Pintos. Some minor mechanical problems such as broken reactor core pins and cracks in the welds of the outer casing of the thermal reactor occurred at 51,000 and 84,000 miles. The pins were replaced, the welds repaired and the test continued. These failures were due to mechanical design problems which could be easily rectified in commercial designs. The interior reactor core and shield and all other hot parts remained in good condition throughout the test.

The average emission levels from replicate emission tests for the TECS III on the 1971 Pinto and a 1971 Chevrolet are given in Table 7. In addition, the results for the modified 1973 Chevrolet with TECS IV tuned to meet different emission levels are included.

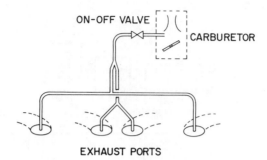

Figure 10. *Exhaust gas recirculation system for 1.6 liter Pinto*

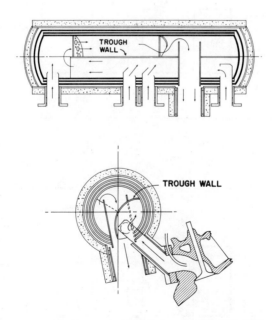

Figure 11. *Type IX thermal reactor for 350 CID Chevrolet*

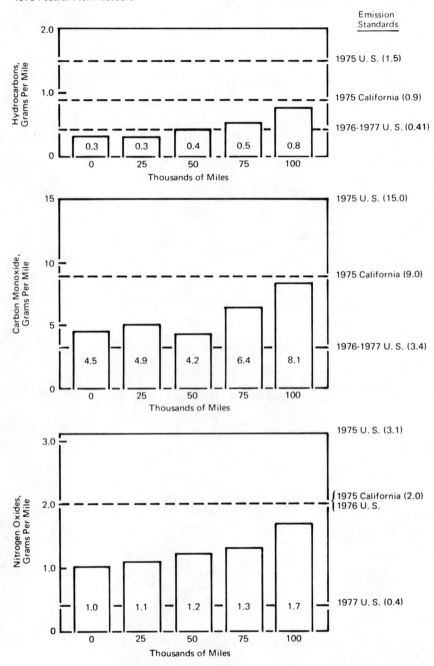

Figure 12. *Exhaust emission levels of TECS-equipped Pinto during 100,000-mile road test*

Table 7

Emission Levels of Vehicles With
TECS III and TECS IV

Vehicle	Control System	Emissions, Grams Per Mile HC	CO	NO$_x$
1971 Pinto 1.6 Liter	TECS III	0.5	5.9	1.3
1971 Chevrolet 350 CID	TECS III	0.4	7.0	1.0
1973 Chevrolet 350 CID	TECS IV	0.4	8.0	1.5
1973 Chevrolet 350 CID	TECS IV	0.4	11.3	2.3
1975 Interim Standards				
U.S. 49-State		1.5	15.0	3.1
California		0.9	9.0	2.0

Fuel Economy. The fuel economy of all vehicles was measured over a 26-mile city-suburban course which included approximately 9 miles of city traffic and 17 miles of suburban roads and interstate highways. The average speed was 30 mph. The test vehicles were fueled from auxiliary tanks located in the trunk and the fuel consumption was determined by weight difference. The emission control vehicles and the standard production vehicles used for comparison were driven in convoy in order to insure that variations in traffic and thus variations in average speed on the course did not influence results. The drivers and the position of the cars were rotated and at least four replicate tests were made. All results are reported as an average of these replicate determinations.

Modifications of carburetion and ignition timing incorporated in the TECS III and TECS IV vehicles have improved fuel economy compared with the earlier emission control systems. In addition, the higher compression ratio engine used with the TECS IV vehicle gave greater improvement in fuel economy than obtained with the TECS III vehicle at similar emission levels. As shown in Table 8, the fuel economy was significantly better with the TECS vehicles than with the 1973 unmodified production vehicle.

As shown in Table 9, the 1.6 liter 1971 Pinto equipped with the TECS III system had an average fuel economy penalty of 6.9% as compared with the unmodified production vehicle. A recent change in the city portion of the driving course to avoid more congested streets explains the greater fuel economy for both cars at the 100,000-mile test point. However, the percent difference in fuel economy remained about the same as for earlier tests. The fuel economy of the TECS III Pinto was 6% better than that measured for two unmodified 1973 Pintos. Minor improvements in ignition timing, carburetion, and, most importantly, a

significant reduction in the pressure drop across the muffler lead trap
when compared with the lead trap incorporated in the TECS II vehicle
resulted in a significant improvement in acceleration time for the TECS
III vehicle. When compared with the unmodified vehicle, the TECS III
car had an 11% increase in wide-open-throttle acceleration time.

Table 8

Fuel Economy of Chevrolets Equipped With
TECS III and TECS IV

	Fuel Economy	
	Miles/Gallon	% Gain*
Unmodified 1973, 350 CID Chevrolet, 8.2:1 C.R.	13.2	–
1971 Chevrolet With TECS III 8.2:1 C.R. Meeting 1975 California Standards	13.8	4.5
1973 Chevrolet With TECS IV, 9.25:1 C.R. Meeting 1975 U.S. Standards	14.6	11
1973 Chevrolet With TECS IV 9.25:1 C.R. Meeting 1975 California Standards	14.1	8

* Compared with 1973

Table 9

Fuel Economy of 1971 Pinto
With and Without TECS III

Vehicles Miles	Fuel Economy, mpg		% Change
	Unmodified Pinto	TECS III Pinto	
0	25.0	22.4	-10.4
25,000	25.3	24.2	- 4.4
50,000	24.0	22.8	- 5.0
75,000	24.9	22.8	- 8.4
100,000*	29.0	27.2	- 6.2
		Average	- 6.9

* Course change

During the course of developing emission control systems based on thermal reactors, a program was undertaken to determine what levels of emission control could be achieved without any form of exhaust after-treatment. The possibility that less stringent standards might be satisfactory for achieving desired air quality gave added impetus for proceeding with this program. The incentive for doing this was the belief that markedly improved fuel economy could be obtained.

A standard 1970 Chevrolet with a 350 CID, V-8 engine was modified by (a) changing cylinder heads to increase the compression ratio to 10.5 to 1 and provide a fast burn chamber, (b) installing head land pistons and rings to reduce the quench zone to reduce unburned hydrocarbons, (c) installing a leaner 4-barrel carburetor, (d) installing a production 1973 intake manifold to provide 1973 production EGR to reduce nitrogen oxides, (e) optimizing camshaft timing, and (f) optimizing ignition timing. The ignition timing was advanced 5° from the manufacturer's specification for a 1970 model and the transmission control spark system was deactivated. The camshaft was advanced about 5° from standard for the 1970 Chevrolet. Carburetor metering was set to give lean operation with air-fuel ratio ranging from 16 to 17:1. No air injection, add-on thermal or catalytic reactors, special high energy ignition system or special choke were used. All components used in the modification were stock parts and easily obtained.

The results to date with only "internal" engine modifications show striking improvements in fuel economy, driveability, and reduced emissions. As shown in Table 10, the emission levels were well below the 1974 emission standards and except for the hydrocarbon levels, were below the 1975 U.S. interim emission standards. The fuel economy, Table 11, was 6% better than a comparable 1967 model and 21% better than a 1973 model as measured in replicate tests on the Du Pont city-suburban road course.

Table 10

Emissions With Only Engine Modifications
No Thermal Or Catalytic Reactor

Vehicle	Emissions, Grams Per Mile 1975 CVS Test Procedure		
	HC	CO	NO_x
1970 Chevrolet 350 CID 10.5:1 C.R.	2.0	6.1	2.2
1974 Standards	3.0	28	3.1
1975 Standards U.S.	1.5	15	3.1

Table 11

Fuel Economy Obtained With Standard and
Modified High Compression Ratio Chevrolet

Vehicle	Fuel Economy	
	Miles/Gallon	% Change vs. 1973
Unmodified 1973 Chevrolet	13.3	–
Modified 1970 Chevrolet		
With 10.5:1 C.R.	16.1	+21

Lead Tolerant Emission Control Systems. Based on the results
presented in the above sections, we believe it is possible to devise
emission control systems which are capable of reducing emission levels
on both large and small vehicles to values which are compatible with
meeting the national ambient air quality standards. Furthermore, these
emission control systems exhibit fuel economy and vehicle performance
characteristics which are better than current production vehicles. These
systems can be used with leaded gasoline and thus provide the motorist
with the benefits of less expensive gasoline and higher compression ratio,
more efficient engines.

If lead continues to be used in gasoline at current levels and it is
prudent to reduce vehicle lead emissions to avoid any potential health
risk, lead traps could be used. The performance of the Du Pont lead
traps is described in the next section.

Du Pont Lead Traps

Design and Performance. The Du Pont muffler lead traps have been
developed to reduce the level of particulate lead material emitted in the
exhaust of vehicles operated on conventional leaded gasoline. The muffler
lead trap, shown in Figure 13, is a single unit, sized and shaped like a
conventional muffler, and located in the same position under the car.
The trap contains a bed of alumina pellets which induces agglomeration
and growth of lead particulate to larger particles. These large lead
particles eventually flake off the alumina pellets and are reentrained in
the exhaust gas stream. The exhaust gas then passes into two parallel
inertial cyclone separators which separate and retain the lead particles
in collection chambers. The collection chambers can be sized to hold all
the exhaust particulate matter collected during 100,000 miles of operation
on conventional leaded gasoline. The Du Pont muffler lead trap has been
used on a wide variety of cars – both current production vehicles as well
as those equipped with advanced emission control systems.

Identical muffler lead traps were installed in place of the exhaust muffler on each of four 1970, 350 CID Chevrolets. These cars were operated on laboratory chassis dynamometers according to the Federal mileage accumulation schedule using gasoline containing 2.2 grams of lead per gallon. During this mileage accumulation, the total lead emissions were measured according to the procedure previously described (12). The average total lead emission rate for each vehicle during the life of the test is summarized in Table 12.

Table 12

Du Pont Muffler Lead Traps
On 1970 Chevrolet

Vehicle	Miles	Lead Emitted, Grams/Mile	% Reduction Due to Trap*
C-82	52,669	0.015	87
C-83	48,735	0.015	87
C-85	50,102	0.015	87
C-130	12,440	0.012	89
Weighted Mean		0.015	88

* Average lead emission rate of production
  Chevrolets is 0.116 g Pb/mile (Ref. 12, Table 1)

In addition to reducing the total amount of lead emitted, the muffler lead traps also reduce the amount of the small, air-suspendible lead particles emitted to the atmosphere. The effectiveness of the lead traps in reducing the emission of lead particles of various sizes at 15,000 and 50,000 miles is shown in Table 13. As expected, the lead traps with their inertial separators were effective in removing almost all of the large particles which for the most part flaked off the walls of the exhaust system. Of special interest is the 68% reduction in the submicron particles. The reduced emission of these small particles is most important since these particles tend to remain airborne and correspond to the size of lead particles found in the atmosphere of urban areas. The efficiency of the traps for capturing these small particles was about the same at 15,000 miles and 50,000 miles. This indicates that the alumina pellets did not deteriorate with mileage. It also indicates that fresh alumina surface is not required which is good, since the pellets tend to become coated with lead salts with extended mileage. High surface area and optimum temperature appear to be much more important factors in trapping the lead.

Table 13

Muffler Lead Trap Performance
On 1970 Chevrolets

| | | Lead Emission Rate, g/Mile | | |
| | | Particle Size, Microns | | |
Vehicle	Mileage	>9	1 to 9	<1.0
Standard Car*		0.038	0.023	0.047
Trap Cars				
C-82	14,500	0.00076	0.0024	0.0177
C-83	3,300	0.00056	0.0031	0.0150
C-85	9,000	0.00074	0.0021	0.0130
Average		0.00069	0.0025	0.0147
% Reduction Due to Trap		98	89	69
C-82	53,600	0.00070	0.0053	0.0165
C-85	51,200	0.00051	0.0029	0.0136
Average		0.00060	0.0041	0.0151
% Reduction Due to Trap		98	82	68

* Table 1 and 2, Ref. 12, average of results from tests of
   5,000 to 55,000 miles

The durability of the muffler lead traps is excellent. Several of the traps were removed from the cars, opened, and examined after 50,000 miles of operation. The alumina pellets were undamaged and except for discoloration showed no deterioration. The pellet bed was not plugged and the bed back pressure did not increase during 50,000 miles of operation. The trap-equipped cars had the same fuel economy and acceleration as cars with conventional exhaust systems.

A lead material balance on car C-82 showed 61% of the lead burned ended up in the trap, 22% in the oil, oil filter, engine, and exhaust pipes, and 12% emitted. Most if not all of the 5% unaccounted for was probably lost in handling the trap. Lead analysis showed the trap gained 15 pounds weight during 50,000 miles of which 9 pounds was lead. The major constituent of the salts in the trap collection chamber was PbBrCl, while the main compound on the pellets was $PbSO_4$. A complete summary of the lead distribution is given in Table 14.

Table 14

Lead Balance of Muffler Lead Trap
From C-82

1970, 350 CID Chevrolet
2.2 g Lead Per Gallon
53,600 Miles
Total Pb Burned = 6,630 Grams

	Lead, Grams	% Lead Burned
Lead Trap		
Cyclones	2,500	37.8
Pellets	1,240	18.7
Scrapings	300	4.5
	4,040	61.0
Emitted	794	11.9
Estimated		
Oil and Oil Filter		11
Engine		3
Manifold and Exhaust Pipes		8
Total Accounted For		94.9

Performance With TECS. Lead traps also can be used in conjunction with thermal reactor based emission control systems. A lead trap of the same design as that shown in Figure 13 was installed on a Pinto in the same location as the conventional muffler on a current production Pinto. The Pinto also was equipped with other components of the TECS III system as shown in Figure 6.

The lead emission rates of the TECS III Pinto and the companion unmodified Pinto were measured during both road and laboratory chassis dynamometer operation during the 100,000-mile test according to procedures previously described (12). The average total lead emission rate for the two vehicles is shown in Table 15. The muffler lead trap system reduced the total amount of lead emitted over the 100,000-mile test by 84%. Only 8% of the lead burned was emitted. There was no deterioration in lead trap performance with mileage. Durability of the trap was excellent. The alumina bed did not plug and back pressure did not increase over the original value during 100,000 miles of operation. There was no deterioration or loss of alumina pellets.

Table 15

Total Lead Emission Rate From 1971 Pinto
With and Without TECS III

Vehicle	Miles	Avg Total Lead Emission Grams/Mile	% Reduction
Unmodified 1971 Pinto	50,000	0.0348	
	101,000	0.0366	
TECS III 1971 Pinto	50,000	0.0063	82
	101,000	0.0060	84

The size distribution of the emitted lead particles from both Pintos was determined at 25,000 and 50,000 miles during the test. The lead emission rates in different particle size categories for the two vehicles are shown in Table 16. Each result is the average of three runs. The cars were driven on a Clayton dynamometer for three Federal test cycles (120 miles) starting from a cold start after an overnight soak at ambient temperatures. The gasoline contained 2.2 grams of lead per gallon. The muffler lead trap was effective, reducing the amount of large particles by better than 97%. It was also effective in reducing the amount of small particles which could be expected to remain airborne for significant lengths of time. The lead trapping system reduced the emission of fine particles below 1.0 microns in diameter by 82%.

Table 16

Size Distribution of Emitted Lead
Particles from TECS Pinto

Particle Size, Microns	Lead Emissions, g/Mile >9.0	1.0 to 9.0	<1.0
Unmodified 1971 Production			
20,000 Miles	0.019	0.018	0.060
51,000 Miles	0.013	0.016	0.088
Average	0.016	0.017	0.074
TECS With Trap			
26,000 Miles	0.0005	0.0016	0.0099
51,000 Miles	0.0003	0.0029	0.0178
Average	0.0004	0.0022	0.0138
% Reduction Due to Trap	97.5	87	82

San Francisco Field Test of Lead Traps. A field test is under way of Du Pont muffler lead traps installed on four cars in San Francisco, California. The test is being conducted in cooperation with the Bay Area Air Pollution Control District, who own and operate the cars. The exhaust muffler on each of two 1971 and two 1972, 6-cylinder, 258 CID Hornets was removed and replaced with a Du Pont muffler lead trap. The traps, designed to last 50,000 miles are sized and shaped like conventional mufflers and installed in the same position.

Every three months, lead emissions are measured with a total filter with the vehicle operated according to the 1968 Federal Test Procedure on a chassis dynamometer. Each car is driven on the dynamometer for a distance of 22.5 miles (25, 137-second, 7-mode cycles) using tank fuel which usually contains about 3.0 grams of lead per gallon. As shown in Table 17, during 1-1/2-years of service, involving 14,000 to 31,000 road miles, the traps reduced the average lead emissions from these cars to 0.0068 gram per mile, equivalent to 4.2% of the fuel lead burned.

Du Pont has been operating two similar cars with conventional mufflers on a programmed chassis dynamometer. Periodically, the lead emissions have been measured using the same driving schedule as the field test cars. The two production cars emitted 0.086 gram Pb per mile or 46% of the lead burned. When the San Francisco cars are compared with the standard cars, it is clear that the traps reduced the lead emitted by 91%.

Table 17

San Francisco Field Test of Du Pont Muffler
Lead Traps on 258 CID, 1971-72 Hornets

Vehicle	Trap Mileage	Average of Total Lead Emissions	
		g Pb/Mile	% of Pb Burned
53	24,455	0.0055	3.6
33	23,902	0.0038	2.2
22	31,515	0.0062	2.9
64	14,124	0.0115	6.8
Average	23,499	0.0068	4.2

Measurement of Lead Emissions. One problem in lead trap development is measuring lead emissions from cars. EPA has stated that emission rates as measured by different investigators vary widely (13). We have pointed out to EPA inconsistencies in their evaluation of vehicle exhaust particulate lead emission measurements and have identified for them reasons for variability of lead emission rates from standard production vehicles (14). These reasons are summarized below.

Lead particulate emissions from cars with standard exhaust systems vary much more erratically than those for cars equipped with lead traps as shown in Figure 14. Lead emissions from standard cars also continue to increase with age as reflected by the finding that the emission rate of a car equipped with a conventional exhaust system was still increasing after 30,000 miles. The lead emission rate of a car equipped with a muffler lead trap showed very little variation and even after 40,000 miles was about the same rate as observed at low mileage. The increase in lead emissions with mileage and the variable lead emission rate of cars with standard exhaust systems is due to the continuous deposition of lead salts on and reentrainment of the lead deposits from the inside surfaces of the exhaust pipe as well as the muffler. This variability of lead emissions does not occur in trap-equipped cars because the cyclones at the end of the trap retain these large particles. Accordingly, lead emissions from trap-equipped cars can be more accurately measured than emissions from uncontrolled cars.

In order to obtain a lead emission baseline for uncontrolled vehicles it is necessary to make measurements over the entire life of several cars using a representative driving cycle. A second approach is to measure the lead emission rate from a large number of cars using the same fuel lead content and driving cycle. The large number is required to obtain a good average since cars have erratic lead emissions even with identical test procedure and similar driving history.

Measurements of lead emission rates of several trap-equipped cars by four organizations showed close agreement. As shown by the data in Table 18, Esso Research & Engineering and Du Pont obtained about the same lead emission rate value from a car equipped with a Du Pont muffler lead trap. For each test value reported by Du Pont and Esso R&E, the car was driven on a dynamometer for one 1972 Federal Test Procedure cycle (7-1/2 miles) from an initial cold start after an overnight soak. The fuel used in these tests contained 2.2 grams of lead per gallon. Esso R&E measured lead emission rate with their sampling system (15) while Du Pont employed the total exhaust filter procedure (12).

A second car, equipped with an advanced gaseous emission control system plus muffler lead traps, was tested by Ethyl Corporation in Detroit and by Du Pont using the 1968 Federal Test Procedure. The gasoline contained 2.2 grams of lead per gallon. Ethyl used the procedure described by Ter Haar, et al (16) to measure lead emissions, while Du Pont employed the total exhaust filter procedure (12). As shown in Table 18, Du Pont and Ethyl obtained almost the same lead emission rate.

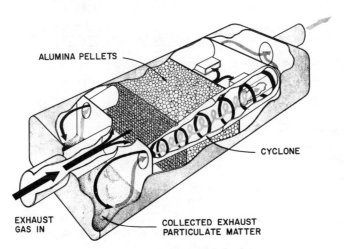

Figure 13.   *Du Pont muffler lead trap*

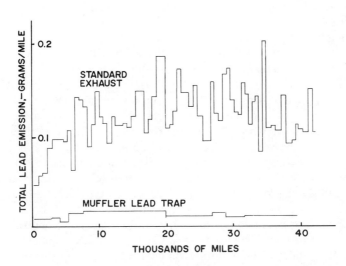

Figure 14.   *Lead emission rate from a car with standard exhaust and a car with a muffler lead trap*

Table 18

Lead Emissions Measured By Different Laboratories

	Test Mode	Lead Emissions, g/Mile	
		Du Pont	Esso R&E
Chevrolet With Lead Trap	CVS-1972 Test Procedure	0.018	0.019

		Dow	Du Pont	Ethyl
Chevrolet With Thermal Reactor, EGR, and Lead Traps	CVS-1974 Test Procedure	0.020*	0.017	-
	1968 Federal Test Procedure	-	0.018	0.021

* Average for six runs, using average Pb content of 2.7 g/gallon

The trap-equipped car tested by Ethyl was also tested by Dow Chemical Company, Midland, Michigan under an EPA contract at the request of EPA. The lead emission rate of this car was measured under a variety of driving cycles consisting of cold and hot start Federal emission test procedure as well as hot continuous 60 mph operation. The value of 0.020 g Pb/mile shown in Table 18 represents an average of six tests by Dow using two fuels, one containing 2.2 and the other 3.2 grams lead per gallon (17). The lead emission rates from this car as measured by these two laboratories and by Du Pont using different driving conditions and different fuel lead content show excellent agreement.

As reported in a letter to EPA (14) there are a number of inadequacies in the Dow measurement of lead emissions from the Du Pont trap-equipped car. Dow measured the weight of total exhaust particulate matter in nine tests with this car and determined lead as a percent of the total for six of these tests. The measurement of total exhaust particulate emission rate by weight is very difficult and subject to serious errors as noted in the Du Pont report to EPA. The total particulate emission rate varied three- or fourfold depending on the type of filter and the sample flow rate used to collect the particulate matter. In addition, three different driving cycles and two fuels with lead content ranging from 2.2 to 3.2 grams of lead per gallon were used. As a consequence of these factors the results of lead emissions rate measured by Dow with the Du Pont car varied from 0.006 to 0.062 gram per mile. However, in spite of these inadequacies, the average total lead emission rate of 0.020 gram lead per mile from the trap-equipped car was in excellent agreement with values reported by Du Pont and others.

Potential of Alternate Emission Control Systems. Non-catalytic emission control systems, such as the thermal reactor-EGR based system, have been developed which are capable of reducing vehicle exhaust emission levels to values which are compatible with meeting the national ambient air quality standards. Thus, the third and final step in the process of determining the best ways of achieving control of automotive air pollution can be considered. This consists of comparing alternate systems, such as catalytic and non-catalytic emission control systems, to determine the most economical means to achieve the desired control of emissions. The final section of this paper provides a critical analysis of the effect of thermal and catalytic control systems on fuel economy.

Effect of Emission Control on Fuel Economy

Because of the energy shortage and the need to reduce gasoline consumption to meet available supplies, attention is being directed toward the effect of automotive emission control devices on automobile fuel economy. This subject has been explored in recent hearings by both houses of Congress as part of their consideration of emergency energy legislation. The Congress is studying the desirability of delaying scheduled changes in emission standards with specific interest in the relationship between the severity of the standards and fuel economy. To assist the Congress in this determination considerable information has been presented in testimony. A number of recent technical publications also have addressed this subject. From a review and analysis of this literature, excellent agreement is found among the various sources as to the deterioration of fuel economy which has taken place as a result of changes already made in cars to reduce exhaust emissions; there is also substantial agreement as to the anticipated effects of imposition of more stringent emission standards presently scheduled for 1975, 1976, and 1977.

Basis for Fuel Economy Comparisons. All comparisons of fuel economy in this report have been made, insofar as possible, at constant vehicle weight, constant engine size, and constant accessory packages. On this basis, the changes in fuel economy reflect primarily only the effects of changes in engine design, carburetion, and ignition timing made to control exhaust emissions. All data were expressed in terms of percent change from 1967 and from 1974. Effects of reduced compression ratios were included since changes in compression ratio were made to reduce octane requirement in anticipation of the use of unleaded gasoline with catalytic control systems. Emission levels are expressed in

terms of measurements made using the constant volume Federal
Emission Test Procedure for 1975 model vehicles.

Change in Fuel Economy From 1967 to 1974. The changes in fuel
economy for model years 1967 through 1974 are shown in Figure 15
according to the data from representatives of Ford, General Motors,
Esso Research & Engineering, the Environmental Protection Agency,
and Du Pont.

The Ford data for model years 1967 to 1973 are averages of fuel
economy changes derived for eight Ford cars weighted to represent the
1967 sales mix (18). The effects of vehicle weight were eliminated and
the observed changes are due only to emission controls and compression
ratio changes. Data for 1974 vehicles are on the same basis and were
presented before the U.S. Senate and House of Representatives (19) (20).
Two points are shown for the 1974 models, one for cars meeting the 1974
U.S. 49-State standards and one for cars meeting the more stringent
1974 California standards.

The General Motors data for model years 1970 to 1974 were taken
from the last figure of the Attachment to the statement presented to the
U.S. Senate Public Works Committee (21). The effects of vehicle weight
were eliminated to show only the effects of emission controls and compression ratio reductions. Based on other data in Reference 21, it is
assumed that the 1970 vehicle was 6% poorer in fuel economy than the
1967 models to tie the data to a 1967 base vehicle. Two points are shown
for the 1974 models, one for the 49-State cars and one for cars meeting
the California standards. The difference between the California and 49-
State results was obtained from the table "Exhaust Emissions vs. Fuel
Economy" in Reference 21.

The data from Esso R&E were obtained in a private communication
and represent the changes in fuel economy each year due to changes in
compression ratios and the application of emission controls (22). The
EPA data show the changes in fuel economy due to emission controls and
compression ratio changes for a 4,500 pound inertia weight vehicle (23).
This size vehicle was chosen to correspond to the weights of the vehicles
used by the other investigators. The Du Pont data from 1970 through
1973 are based on measurements of six representative standard size
sedans purchased each year since 1970 (11). The effects of changes in
vehicle weight were eliminated. Based on a consensus of data from other
sources, the 1970 vehicles were assumed to be 4% poorer in fuel economy
than the 1967 vehicles.

As shown in Figure 15 all five investigators show a decrease in fuel
economy from 1967 to 1973. The average fuel economy loss for the 1973
cars relative to that of the 1967 cars is 14.1% with a range in values from
13.5 to 14.5%. This value is further confirmed by data presented by

Chrysler which showed a 14% decrease for a typical 1973 car (24). Agreement in the overall loss for 1973 models is excellent considering the differences in types of cars and driving cycles used to measure fuel economy. The agreement on the fuel economy loss for the 1974 models is not as good as was the case for the 1973 cars. An average loss of 14.2% compared with 1967 models was obtained for the 1974 cars which is about the same as for the 1973 models; however, the values range from 10 to 18%. An average loss of 17.7% relative to 1967 was obtained for cars meeting the more stringent 1974 California standards.

The loss in economy due to the reduction of compression ratio by approximately one unit from the 1970 to 1974 models has been estimated to range from a low of 2% to a high of 7% by the different investigators. If an average value of 5% due to compression ratio reduction is assumed, then the loss due to engine modification for emission control is 9%. In addition to the 14% loss in fuel economy, there also has been a loss of about 10% in vehicle acceleration performance due to emission controls and reduced compression ratio. If vehicle acceleration performance had been maintained constant, the fuel economy loss would have been approximately 20%.

Effect of Proposed Catalytic Control Systems. The projected changes in fuel economy of cars equipped with catalytic emission control systems proposed to meet various levels of exhaust emission standards are shown in Figure 16. Data presented by representatives of Ford, General Motors, Esso Research & Engineering, Chrysler, the EPA, and Du Pont were used in preparing this figure.

The Ford data were taken from the figures accompanying the statements to the U.S. Senate and House of Representatives (19) (20). The Chrysler data were taken from an SAE paper by Huebner (24) and a statement by Chrysler to the U.S. House of Representatives (25). The Esso R&E data were presented to the U.S. House of Representatives by Russell Train, Administrator of the EPA, in support of his statements (26). The EPA data were taken from the SAE paper by Austin and Hellman (23).

The General Motors data were derived from several sources. The data point for vehicles meeting the 1975 U.S. interim standards was taken from the colloquy following the statement presented by GM to the U.S. House of Representatives (27). In the prepared statement it is stated that the fuel economy of the 1975 vehicles would be 13% better than the 1974 vehicles on a sales weighted basis. In the oral testimony, it was stated that some of the 13% gain was due to the increased sales of smaller cars and that only a 10% gain should be attributed to the use of the catalytic control systems and the accompanying engine adjustments. It was also stated that meeting standards more stringent than the U.S. 1975 interim levels would cause fuel economy to become progressively poorer

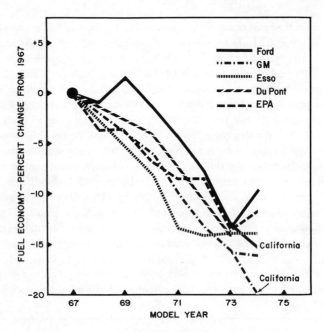

Figure 15. Effect of reduced compression ratio and emission controls on the fuel economy of 1968–1974 model cars

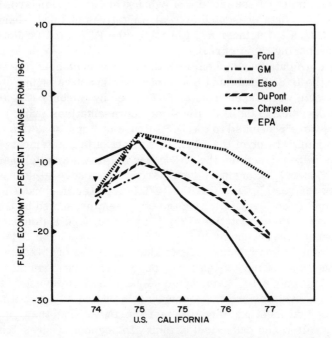

Figure 16. Effect of proposed catalytic emission control systems on fuel economy

but the expected decrease was not quantified. A quantitative measure of these changes was obtained from data in the table "Exhaust Emissions vs. Fuel Economy" from the Attachment to the earlier GM statement to the U. S. Senate (21). In this table, the fuel economy changes were given for one car as different emission control systems were installed to meet a variety of emission standards ranging from the current 1974 levels to the stringent 1977 levels. The fuel economy of this particular car was very responsive to changes made in the various adjustments and these data appear to have formed the basis for General Motors' earlier comments that the fuel economy of the 1975 models could be improved by 20% relative to the 1974 models. Since this particular car had an improvement of 20% in fuel economy compared with 1974 while all General Motors cars apparently will improve only 10% when meeting the 1975 U. S. interim standards, it was assumed for the purposes of this report that the changes in economy corresponding to other standards would be similarly magnified. Accordingly, the fuel economy changes at the various emission levels relative to the 1974 production car were divided by two and the resulting fuel economy changes are shown for the 1975 California, 1976, and 1977 levels.

The Du Pont data for the 1975 U. S. and California interim emission levels were estimates based on the extent by which the carburetion and spark timing could be changed to improve economy while still meeting the required NO_x levels and allowing the hydrocarbon and carbon monoxide levels to be controlled by a catalytic reactor. The data for the 1976 and 1977 levels are based on fuel economy tests conducted with 1972 Pintos equipped with 2. 0 liter engines and automatic transmissions. The emission control vehicles were equipped with low thermal inertia manifolds, an oxidizing catalytic reactor for 1976, and both oxidizing and reducing catalytic reactors for 1977. Low mileage emission levels were well below one-half of the required levels for 1976 and 1977. Extended mileage tests have not yet been conducted.

1975 U. S. interim standards. Estimates of the fuel economy improvements for cars meeting the 1975 U. S. standards compared with the 1974 vehicles range from 3 to 10%. Ford and Chrysler estimated an increase of 3%, Du Pont 4%, Esso R&E 8%, and General Motors 10%. The large discrepancy between the estimates appears to be due in the main to the different starting points for the 1974 vehicles. The loss relative to 1967 models for the 1975 U. S. cars ranges from 6 to 7% compared with the 1967 cars according to Ford, GM, and Esso R&E. Du Pont and Chrysler estimate the loss relative to 1967 to be 10 and 12%, respectively.

Compared with 1967, the fuel economy loss to meet the 1975 U. S. interim standards is 8% when values for the five organizations are averaged. Since the average loss for the 1974 models is approximately 14%,

there would appear to be an average improvement for meeting the 1975
U. S. interim standards of 6% compared with the 1974 models. As discussed previously, there was a loss of approximately 9% for the 1974
models compared with the 1967 models due to the emission controls only.
Thus, approximately two-thirds of this loss apparently will be recovered.
However, the remaining loss of 3% plus the 5% loss due to the lowered
compression ratio will remain.

1975 California interim standards. All sources agree that meeting
the more stringent 1975 California standards will cause a loss in fuel
economy compared with the models meeting the 1975 U.S. standards.
The estimates of this loss are 1% by Esso R&E, 2% by GM and Du Pont,
and 8% by Ford giving an average loss of 3%. When compared with the
1967 models, the average loss in fuel economy for the 1975 California
cars would be 11%. Thus, the 1975 California cars would be 3% better
in fuel economy than present 1974 models.

1976 Standards. Again all sources agree that a further loss in fuel
economy would be incurred to meet 1976 standards when compared with
1975 California cars. Estimates are 1% by Esso R&E, 4% by GM and
Du Pont, and 5% by Ford. The loss compared with the 1967 cars is 8%
by Esso R&E, 13% by General Motors, 16% by Du Pont, and 20% by Ford.
Data from the EPA report (23) shows a loss of 13.7% for 36 prototype,
4,500-pound vehicles built primarily by the automobile manufacturers to
meet the 1976 standards. The average loss according to these five
sources is 14% for cars meeting the 1976 standards compared with 1967
models. Thus, these cars would be apparently equivalent in fuel economy
to present 1974 vehicles. However, the Esso R&E estimates of the fuel
economy loss are much lower and appear to be out of line with the estimates from the four other sources. If the Esso R&E data are excluded,
the average loss for 1976 model cars would be 16% compared with the
1967 cars. In this case, the 1976 cars are 2% poorer in fuel economy
than the 1974 cars.

1977 Standards. Cars meeting the 1977 standards will suffer a large
loss in fuel economy to meet the very low NO_x level of 0.4 gram per mile.
The estimates compared with the 1967 models are 12% by Esso R&E,
20.5% by GM, 21% by Du Pont, and 30% by Ford. The average loss from
these four sources is 21% for 1977 models compared with 1967 models.
Again, the data from Esso R&E would appear to be out of line compared
with the other sources. If the Esso R&E data are excluded the average
loss for 1977 cars compared with 1967 models is 24% or these cars
would be 10% poorer than the present 1974 models.

Overall estimate of the fuel economy of catalytic systems. Cars equipped with catalytic reactors and low compression ratio engines and built to meet the 1975 U. S. or California interim standards will have improved fuel economy compared with 1974 models based on the data from the six sources cited in this report. Cars meeting the more stringent 1976 or 1977 statutory standards will have poorer fuel economy than 1974 cars. In all cases the fuel economy will be poorer than that of 1967 cars due to lower compression ratios and the imposition of the emission controls. The average data are summarized in Table 19.

Table 19

Effect of Catalytic Emission
Control System on Fuel Economy

Low Compression Ratio Engines (8.3:1)

	Emission Levels Grams/Mile			Change in Fuel Economy, %	
Emission Standards	HC	CO	NO$_x$	Compared With 1974 Models	Compared With 1967 Models
1975 U. S. Interim	1.5	15	3.1	+ 6	− 8
1975 California Interim	0.9	9.0	2.0	+ 3	−11
1976 Statutory	0.41	3.4	2.0	− 2	−16
1977 Statutory	0.41	3.4	0.4	−10	−24

Effect of Thermal Reactor Control Systems. An alternative way to achieve low exhaust emission levels would be to use control systems based on thermal reactors. Because these systems are not sensitive to leaded gasoline, the compression ratio could be raised to improve fuel economy. Changes in fuel economy of cars equipped with thermal reactor and exhaust gas recirculation emission control systems to meet various levels of exhaust emission standards are shown in Figure 17. These vehicles were equipped with high compression ratio engines designed to run on leaded fuel. Data from Esso R&E and Du Pont were used. Comparable data are not available from the automobile manufacturers since all of their efforts appear to be directed toward using catalytic reactors as the only means to achieve the very stringent 1977 emission standards.

Esso Research & Engineering data. The Esso R&E results were presented together with the EPA statement to the U. S. House of Representatives (26). According to a private communication from Esso R&E the result for the 1974 vehicles was calculated based on the improvement in fuel economy which could be expected if the compression ratios of 1974 vehicles were restored to 1970 levels. The data points for the 1975 U. S. interim standards and the 1976 and 1977 standards according to Esso R&E were based on tests of experimental vehicles. These results probably were obtained with Synchrothermal and the RAM reactors developed by Esso R&E (28) (29).

Du Pont data. The data from Du Pont are based on the results with the Total Emission Control Systems (TECS) developed by Du Pont. These TECS vehicles were equipped with thermal reactors, EGR systems, and lead traps. Carburetion and spark advance modifications were made to obtain the desired degree of emission control and to minimize fuel economy penalties. Pertinent data from References 10 and 11 and additional data from this paper are summarized in Table 20. Many of these cars were at low mileage, such as 2,000 miles, but some data, such as for the 1971 Pinto in Line 4, were results from durability tests of 20,000 to 100,000 miles in duration.

Table 20

Fuel Economy of Cars Equipped with Various
Du Pont Thermal Emission Control Systems

Base Vehicle	Control System	Emission Levels Grams Per Mile			Fuel Economy Change, % Relative to 1967
		HC	CO	NO_x	
1970 Chevrolet 350	Engine Modification	2.2	8.0	2.0	+ 6
1973 Chevrolet 350	TECS IV	0.4	11.3	2.3	− 3
1973 Chevrolet 350	TECS IV	0.4	8.0	1.5	− 7
1971 Pinto 1.6 Liter	TECS III	0.4	5.0	1.2	− 8
1970 Chevrolet 350	TECS I	0.4	11.0	1.1	−18
1971 Chevrolet 350	TECS III	0.4	7.0	1.0	− 8
1971 Pinto 1.6 Liter	TECS II	0.24	5.0	0.61	−24
1970 Chevrolet 350	TECS II	0.16	6.6	0.57	−30
1971 Chevrolet 350	TECS III	0.07	2.54	0.49	−25

Results shown on the first line of Table 20 were obtained with the 1970 Chevrolet modified to meet the 1974 emission standards and obtain maximum fuel economy as discussed earlier. The emission levels were well below the 1974 standards and except for the hydrocarbon levels, were below the 1975 U.S. interim standards. The fuel economy was 6% better than a comparable 1967 model and 21% better than a 1974 model as measured on the Du Pont city-suburban road course. Thus, it appears feasible to obtain significantly improved fuel economy while meeting the 1974 standards if the compression ratio is raised and fuel economy optimized.

The Du Pont points for the 1975 U. S. and California interim standards in Table 20 were based on the results described in this paper and obtained with TECS IV. The systems employed and the results obtained with the Pinto and Chevrolets in Table 20 have been described earlier in this paper. All of these recent vehicles were designed to meet the 1975 California interim emission standards with minimum fuel economy penalty. The losses ranged from only 7 to 9% relative to that of the 1967 models. The Pinto is a much lighter car equipped with a 4-cylinder engine but the percent change in fuel economy for the TECS vehicle when compared with a production model was the same as that obtained with the standard size sedans. The TECS I vehicles, shown in Line 5, were developed in 1970 with the only consideration being meeting the emission standards. With this lack of emphasis on fuel economy, the loss for this early design was 18% relative to the 1967 vehicles.

The fuel economy loss for Du Pont thermal reactor systems meeting the 1976 standards is based on the results obtained with the vehicle shown in the last line of Table 20. This vehicle handily met the hydrocarbon and carbon monoxide levels for 1976 but, because the vehicle was originally targeted for the 1977 standards, the nitrogen oxides level was well below the 1976 standard. As a result, the fuel economy loss of 25% with this vehicle relative to 1967 models is greater than if the NO_x level had been relaxed to approach the 1976 standard of 2. 0 grams per mile. It is estimated that a fuel economy loss of 20% compared with 1967 would have been obtained if the system were readjusted to give a NO_x level of about 1. 5 grams per mile. Fuel economy values from Table 20 for the various Du Pont TECS cars are superimposed on the trend line in Figure 17 to illustrate how the Du Pont trend line was developed.

Du Pont has not developed a thermal reactor-EGR system which meets the 1977 standards. The very low NO_x level of 0. 4 gram per mile imposes unacceptable fuel economy and driveability penalties. An estimate of the fuel economy penalty incurred in meeting the 1977 standards can be obtained by examination of the data shown in Table 20. Extrapolation of the results gives an estimate of a fuel economy loss approaching 30% compared with 1967 vehicles to meet the 1977 standards.

Comparison of Esso R&E and Du Pont data. The trend lines shown in Figure 17 for the Esso R&E and Du Pont data show significant discrepancies for vehicles meeting the 1974 and 1975 interim standards but good agreement for vehicles meeting the 1976 and 1977 standards. For the 1974 vehicles, the Esso R&E data give credit for only the increase in compression ratio. No credit is given for the additional changes to improve fuel economy such as were made in the Du Pont vehicle. On the other hand, perhaps the vehicle chosen for the Du Pont studies may have been particularly responsive to the changes made and such improvements in fuel economy might not be realized in all models. Also, the compression ratio was 10.5:1 which is well above the 1970 average value of 9.4. For the purposes of this report, it is reasonable to average the two estimates of an improvement of 6% by Du Pont and a loss of 7% by Esso R&E compared with 1967 models. When this is done a fuel economy loss of 1% compared with 1967 vehicles but a gain of 13% compared with 1974 vehicles results for vehicles meeting the 1974 standards but optimized for maximum fuel economy with high octane gasoline.

For the 1975 interim standards, either U.S. or California, the Esso R&E data appear to be based on the results they obtained several years ago with their Synchrothermal reactor system (28). This system used an early version of the Du Pont thermal reactors and the main thrust of the program was to demonstrate low emission levels without full regard for maximizing fuel economy. The Du Pont data were developed recently with improved thermal reactors and with the goal of meeting specific emission standards with a minimum fuel economy penalty. As a result the Du Pont data appear more representative of the fuel economy losses to be expected with thermal reactor systems. These results show an improvement of 11% at the 1975 U.S. interim standards and 6% at the 1975 California interim standards compared with 1974 vehicles equipped with thermal systems optimized for maximum fuel economy with high octane gasoline. These vehicles show losses of 3% and 8% compared with 1967 models at the 1975 U.S. and California interim standards, respectively.

The Esso R&E and Du Pont data are in close agreement for the 1976 and 1977 standards. To be conservative, the lower Du Pont values will be assumed as more representative. These data show a loss of 20% at the 1976 standard and about 30% for 1977 compared with 1967 models for vehicles equipped with thermal systems and operated on high octane leaded gasoline.

Overall estimate of the fuel economy of thermal reactor systems. Cars equipped with thermal systems and high compression ratio engines, optimized for maximum fuel economy, and built to meet the 1974, the 1975 U.S., and the 1975 California interim standards appear to have better fuel economy than 1974 models. Cars built to meet the more stringent 1976 and 1977 standards probably will have poorer fuel economy than the 1974 models. In all cases, the fuel economy will be poorer than the 1967 models due to the imposition of the emission controls. The data are summarized in Table 21.

Table 21

Fuel Economy of Thermal Reactor-EGR Emission
Control Systems Meeting Various Standards

High Compression Ratio Engines (9.4:1)

| Emission Standards | Emission Levels Grams Per Mile | | | Change in Fuel Economy, % | |
	HC	CO	NO$_x$	Compared With 1974 Models	Compared With 1967 Models
1974	3.0	28.0	3.1	+13	− 1
1975 U.S. Interim	1.5	15.0	3.1	+11	− 3
1975 California Interim	0.9	9.0	2.0	+ 6	− 8
1976 Statutory	0.41	3.4	2.0	− 6	−20
1977 Statutory	0.41	3.4	0.4	−16	−30

Comparison of Catalytic and Non-Catalytic Systems.   Average fuel
economy penalties associated with meeting various levels of emission
standards are plotted in Figure 18.   The values for the catalytic systems
operated with low compression ratio engines are taken from Table 19.
The total range of the estimates in fuel economy taken from Figure 16
also is shown by the crosshatched area.   The fuel economy values for the
thermal or non-catalytic systems operated with high compression ratio
engines are taken from Table 21.

At the 1974 standards there is a clear cut advantage for the use of
higher compression ratios than now used for the 1974 models.   If future
cars were built to meet the current 1974 emission standards but with
higher compression engines and optimized fuel economy an immediate
improvement of about 13% in fuel economy compared with the current
1974 models could be realized.   The fuel economy of these cars could
approach that of the uncontrolled 1967 models except for those losses due
to increased vehicle weight and additional accessories.

The fuel economy of vehicles meeting the 1975 U.S. interim standards
equipped with either high compression ratio non-catalytic or low compression ratio catalytic control systems, will be better than that of the
1974 models.   Fuel economy would be increased 11% relative to the 1974
models by raising the compression ratio to 9.4:1 (the average level for
1970), using a thermal reactor emission control system, and optimizing
fuel economy.   Fuel economy would be increased 6% compared with 1974
models by using a catalytic emissions control system, retaining the
present-day low compression ratio engines and optimizing for fuel
economy.   For the 1975 California interim standards there would be a
smaller fuel economy improvement for both systems compared with the
1974 models.   The high compression ratio thermal system would show a

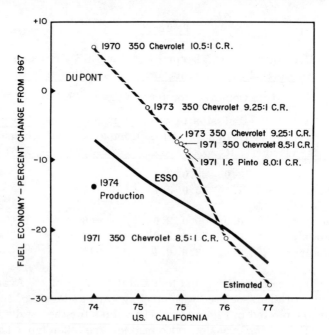

Figure 17.   Effect of thermal emission control systems on fuel
economy

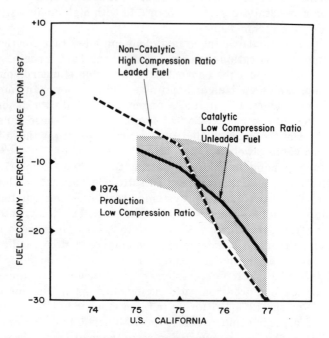

Figure 18.   Comparison of the effect of catalytic and thermal
emission control systems on fuel economy

6% improvement and the low compression ratio catalytic system a 3%
improvement compared with 1974 models.

When emission standards are more stringent than the 1975 California
standards, fuel economy will be worse than that of 1974 models. With the
1976 standards, the fuel economy for the catalytic system is 2% poorer
than 1974 and the thermal system is 6% poorer. With the 1977 standards,
the fuel economy loss is 10% for the catalytic systems and at least 16%
for the thermal systems compared with 1974 models. Compared with
1967 models the fuel economy losses range from 24 to 30% at the 1977
standards.

Fuel Economy Loss to Operate on 91 Octane Unleaded Gasoline. The
foregoing analysis assumes that the 1975 and later model cars equipped
with catalytic systems will operate satisfactorily on the 91 octane number
unleaded gasoline which is required to be made available starting in July,
1974. However, it is well known that many 1973 and 1974 models knock
on 91 octane number unleaded gasoline (30). If the 1975 cars will not
operate knock-free on 91 octane, then further reduction in octane requirements will be required with additional losses in fuel economy. Alterna-
tively, unleaded fuel of greater than 91 octane number will be required
for some of these cars and refining to produce this higher octane number
unleaded gasoline will consume additional crude. Thus, additional fuel
economy penalties not included in the previous sections will occur for the
1975 cars equipped with catalytic systems.

The imposition of emission controls in 1968 resulted in a small fuel
economy loss but had no significant effect on the octane requirement of
vehicles. However, starting in 1971 changes have been made in engines
to reduce octane requirements to permit the use of 91 octane number unleaded gasoline because of the planned installation of catalytic emission
control systems. These octane requirement reductions were accomplished by a reduction in compression ratio, changes in spark and valve
timing, changes in carburetion and induction system design, changes in
combustion chamber design, etc.

The overall relationship between reduced octane requirement and
fuel economy loss can be determined by relating (a) the data from the
Coordinating Research Council (CRC) on octane requirements for vehicles
and (b) the fuel economy losses developed in this report. Each year the
CRC publishes the results of an extensive survey conducted to determine
the octane quality required to avoid knock for the cars produced that
model year (30). The octane numbers of gasolines required to avoid
knock for 10, 50, and 90% of all of the cars for model years 1963 to 1973
are plotted in Figure 19 as a function of the average fuel economy loss
for the corresponding model year as derived from Figure 15. The 1963
to 1967 data are plotted as one average point since little change in octane

requirements occurred and no change in fuel economy was assumed.

Least square regression lines are shown for the years 1970 to 1973 when octane requirements were reduced. The relation between octane requirement and fuel economy is quite good with correlation coefficients exceeding 0.9 in all three cases. As shown, reductions in octane requirements have been accompanied by poorer fuel economy. The slope of the regression lines indicates about 1.4% loss in fuel economy for each one octane number reduction in requirement. This slope is essentially the same as the slope of 1.5% per octane number derived when only one variable such as compression ratio or spark advance was used to reduce octane requirement and the resulting change in fuel economy was determined (31).

Although octane requirements have been reduced substantially since 1970 the CRC data, in Table 22, show that many 1973 models will not operate knock-free on 91 octane number gasolines. The CRC requirements were obtained with leaded rating fuels which are similar to current commercial leaded gasolines and also with unleaded rating fuels which are prototype blends of the average unleaded gasolines anticipated to be made available in the future. However, the CRC data were obtained on cars which had been driven on leaded gasoline and not unleaded gasoline as is projected to be the case with 1975 model cars equipped with catalytic control systems. A recent study by the CRC has shown that the octane requirement increase caused by combustion chamber deposits from unleaded fuels is 2 octane numbers greater than when leaded fuels are used (32). Thus, the octane requirements of the 1973 model cars would have been 2 octane numbers higher if the cars had been operated exclusively on unleaded gasoline. The 1973 CRC octane requirements determined with unleaded rating fuels have been increased by 2 numbers to account for this fact as shown in the last column of Table 22.

Based on work by Du Pont, it appears likely that 1974 models will have equilibrium octane requirements similar to those of 1973 models (33). If 1975 models also have similar octane requirements, only 50% of the 1975 models can be expected to operate knock-free on 91 octane number unleaded gasoline as shown in the last column of Table 22. For pre-1975 models this failure of some cars to operate on 91 octane number gasoline has not been important since the owners use higher octane leaded regular or premium gasolines. However, 1975 cars equipped with catalytic reactors will require unleaded gasoline and the motorist will no longer have the option of using the higher octane leaded fuels.

Traditionally, premium gasolines have satisfied 90% of the premium requirement cars based on past CRC octane requirement studies (30). The 1975 and later models will require the same degree of satisfaction with unleaded gasolines because as mentioned above they cannot use

Table 22

Research Octane Requirements

Percent Of Cars Satisfied	CRC Data 1973 Cars Operated On Leaded Fuel		Corrected For Operation On Unleaded Fuel
	Leaded Rating Fuels	Unleaded Rating Fuels	
10	82.6	83.1	85.1
20	83.9	85.1	87.1
40	86.6	88.1	90.1
50	87.9	89.1	91.1
60	88.7	90.2	92.2
80	90.5	93.0	95.0
90	92.7	95.0	97.0

higher octane leaded gasolines. Thus, as shown in the last column of Table 22, the 1975 model cars will require 97 octane unleaded gasoline to achieve 90% satisfaction.

There are two possible solutions to the problem of providing satisfactory operation on unleaded gasoline. Either vehicle octane requirements can be further reduced so that virtually all new cars will operate on 91 octane number unleaded gasoline or higher octane number unleaded gasoline can be made available. If octane requirements are reduced then further losses in fuel economy will occur. As shown in Table 22, if 90% of the cars are to be satisfied with 91 octane number, then a reduction of 6 octane numbers in the requirements would be needed. A reduction of 6 octane numbers would cause a decrease in fuel economy of 9% based either on earlier work (31) or the relationships developed in Figure 19.

A less wasteful solution to knock-free operation on 91 octane number unleaded gasoline would be to lower the octane requirements of the 50% of the cars that are above 91 octane number. This reduction could be done by new car dealers or service station operators who probably would retard the spark to provide satisfactory operation of those cars on 91 octane number unleaded gasoline. However, retarding the spark will cause a reduction in fuel economy. Based on the octane requirement distribution shown in Table 22 and the relationship between octane reduction and fuel economy loss from Reference 31 or as shown in Figure 19, a reduction in fuel economy of 6% for cars above 91 octane number would occur if their requirements were reduced to 91 octane number. This fuel economy reduction is equivalent to a 3% reduction in fuel economy for all cars.

Higher octane unleaded gasoline also could be supplied. At least two variations on this problem are possible. First, all of the unleaded gaso-

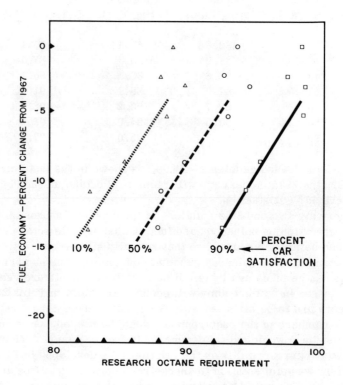

Figure 19. *Effect of octane requirement reduction on fuel economy loss, based on 1967–1973 cars*

line could be provided with a quality high enough to satisfy most of the cars produced. If 90% of the cars are to operate knock-free then the unleaded octane number would have to be increased to approximately 97 as shown in Table 22. However, it has been estimated that from 2 to 6% more crude oil would be needed to produce 97 octane number unleaded gasoline rather than 91 octane number gasoline (34) (35). Therefore, increasing the unleaded gasoline octane number to 97 would use 2 to 6% more crude and would be equivalent to a 4 to 12% decrease in fuel economy.

A second less wasteful solution would be to market two grades of unleaded fuel such as 91 and 97 octane for the cars that would knock with 91 octane number. In this case only 50% of the gasoline would require more crude and the additional crude penalty would be only 1 to 3% depending on the extent of the crude penalty required to produce it. This crude penalty is equivalent to a fuel economy loss of 2 to 6%. This second solution is not practical in the near future because it would require service stations to market two grades of unleaded gasoline in addition to the two grades of leaded gasoline they are already marketing for the almost 100 million vehicles now on the road. It is not likely that the additional tanks and dispensing pumps could be installed except over a period of several years.

Some investigators have claimed (35) (36) that the production of higher octane unleaded gasoline to satisfy the high requirement cars is not wasteful because these high requirement vehicles have better fuel economy and the improved fuel economy more than compensates for the additional crude used. This contention is probably true if the reasons for the high requirements are due to deliberate engineering changes made to improve fuel economy such as increased compression ratio or advanced spark. However, many of the causes for high octane requirements are due to engine variations such as higher intake air temperatures, poor heat transfer to the coolant, etc. which may not improve fuel economy and in some cases may make it poorer.

From this analysis it appears that many 1975 cars will not operate knock-free on the 91 octane number unleaded gasoline which is scheduled to be made available starting in July, 1974. Either vehicle modifications to reduce the octane requirements so that all cars will operate on 91 octane number or the production of higher octane gasoline will entail a penalty. This additional penalty is estimated to range from 2 to 6% with a reasonable value being a 4% loss in fuel economy of 1975 cars compared with 1974 cars. This 4% loss will reduce the fuel economy improvements cited in the previous section for cars equipped with catalytic reactors and low compression ratio engines. This additional 4% loss in fuel economy for cars equipped with catalytic reactors and low compression ratio engines is shown in Table 23 and compared with the loss due to cars with

thermal reactors and high compression ratio engines. From this data
it appears that the thermal reactor high compression ratio systems
provide better fuel economy in meeting the interim 1975 standards while
the catalytic systems are better in meeting the 1977 statutory standards.

Table 23

Effect of Catalytic and Thermal
Emission Control Systems on Fuel Economy

Emission Standards	Loss In Fuel Economy, Percent Compared With 1967 Models	
	Catalytic, Low Compression Ratio	Thermal, High Compression Ratio
1974 U.S.	14*	1**
1975 U.S. Interim	12	4
1975 California Interim	15	8
1976 Statutory	20	20
1977 Statutory	28	30

* Current Production
** No Thermal Reactor

Assessment of Emission Control and Fuel Economy. Fuel economy has
deteriorated due to the imposition of emission control systems and the
reduction of octane requirements by means of reduced compression
ratios and retarded spark timing. This loss could be essentially eliminated and the current 1974 emission standards could still be met if compression ratios were returned to pre-1971 levels and other engine parameters adjusted to maximize fuel economy. The use of advanced emission control systems such as catalytic or thermal reactors will allow
fuel economy to be improved compared with current 1974 vehicles if the
emission standards are not too stringent. The thermal systems because
of their ability to operate with higher octane leaded gasoline offer the
potential for greater improvement in fuel economy than catalytic systems
which would operate on lower octane unleaded gasoline at emission levels
as low as the 1975 California interim standards. If emission standards
as stringent as those mandated by the 1970 Amendments to the Clean Air
Act must be met, the fuel economy penalty will be much larger than
already encountered with today's 1974 vehicles with either catalytic or
thermal control systems.

Summary and Conclusions

Studies of the relationships between vehicle emission rates, traffic density, and ambient air quality levels suggest that a vehicle carbon monoxide emission standard of approximately 26 grams per mile would be sufficient to achieve the ambient air quality standard for carbon monoxide even in the most heavily congested urban areas. Studies by others suggest the hydrocarbon and nitrogen oxide standards mandated by the 1970 Amendments to the Clean Air Act are too stringent by a factor of 3 to 5. In light of these findings, it appears that the 1975 interim emission standards are more stringent than required to meet the ambient air quality standards for automotive related pollutants in most if not all urban areas. The current 1974 emission standards appear to be sufficient for most urban areas and certainly for the remainder of the country.

If automotive emission standards were established at levels indicated to be sufficient to meet the air quality standards, several alternative emission control systems could be employed. Modifications of existing engines will reduce emissions to levels approaching the 1975 interim standards and the use of thermal reactors and exhaust gas recirculation will allow the 1975 interim California standards to be met handily. These control systems are compatible with the continued use of leaded gasoline. Thus, the efficiency of the engine and the fuel economy can be improved by increasing the compression ratio and optimizing the spark advance. Should a need be demonstrated to reduce the lead emissions from vehicles, lead particulate traps have been demonstrated to be a practical and economical means to reduce lead emissions by 70 to 90%.

The improvements in fuel economy attainable with thermal emission control systems used with high compression ratio engines and current leaded gasolines are greater than those attainable with catalytic systems used with low compression ratio engines and unleaded gasolines at emission levels as low as the 1975 interim California standards. At the lower emission standards mandated by the 1970 Amendments to the Clean Air Act the catalytic systems would probably have less fuel economy penalty than the thermal systems. However, at these very stringent standards both systems would have much poorer fuel economy than either pre-emission control or current vehicles.

In view of the critical energy supply problem facing this country for the foreseeable future, these findings call for a re-examination of current strategies for control of automotive related air pollution. Those strategies which are capable of achieving the air quality goals with the least consumption of energy should receive maximum priority.

Acknowledgments

The authors wish to acknowledge the contributions of W. G. Kunz, Jr., I. T. Rosenlund, R. D. Snee, W. S. Vilda, and J. Zelson in the development of the information and vehicles discussed in this paper.

Literature Cited

1. Cumulative Regulatory Effects on the Cost of Automotive Transportation - RECAT, Prepared for the Office of Science and Technology, February 1972.

2. Heuss, J. M., Nebel, G. J., and Collucci, J. M., "National Air Quality Standards for Automotive Pollutants - A Critical Review, J. Air Poll. Control Assoc., 21, 535-44.

3. Pierrard, J. M., Snee, R. D., and Zelson, J., "A New Method of Determining Automotive Emission Standards," Presented at National Meeting, Air Pollution Control Association, June 1973, Chicago, Illinois.

4. Pierrard, J. M. and Snee, R. D., "Relating Automotive Emissions and Urban Air Quality," Prepared for Presentation at National Meeting of Air Pollution Control Association, June 1974, Denver, Colorado.

5. A Study of Emissions from Light Duty Vehicles in Six Cities, U. S. EPA, APTD 1497, March 1973.

6. "Transportation Controls to Reduce Motor Vehicle Emissions in Boston, Mass." U. S. EPA, APTD 1442, December 1972.

   "Transportation Control Strategy Development for the Denver Metropolitan Area." U. S. EPA, APTD 1368, December 1972.

   "Transportation Controls to Reduce Motor Vehicle Emissions in Minneapolis and St. Paul, Minn." U. S. EPA, APTD 1447, December 1972.

   "Transportation Control Strategies for the State Implementation Plan City of Philadelphia." U. S. EPA, APTD 1370, February 1973.

   "Transportation Controls to Reduce Motor Vehicle Emissions in Pittsburgh, Pa." U. S. EPA, APTD 1446, December 1972.

   "Transportation Controls to Reduce Motor Vehicle Emissions in Spokane, Wash." U. S. EPA, APTD 1448, December 1972.

7. "An Interim Report on Motor Vehicle Emission Estimation." U. S. EPA, October 1972.

8. "A Critique of 1975-76 Federal Automobile Emission Standards for Hydrocarbons and Oxides of Nitrogen," Panel on Emission Standards and Panel on Atmospheric Chemistry, Committee on Motor Vehicle Emissions, National Academy of Sciences, May 1973, Washington, D. C.

9. Ruckelshaus, W. D., "Decision of the Administration on Remand from the U. S. Court of Appeals for the District of Columbia Circuit," April 11, 1973, and Accompanying Remarks Given in Press Release.

10. Cantwell, E. N., Hoffman, R. A., Rosenlund, I. T., and Ross, S. W., "A Systems Approach to Vehicle Emission Control," Paper No. 720510, Presented at the Society of Automotive Engineers National Automobile Engineering Meeting, Detroit, Michigan, May 22-26, 1972.

11. Cantwell, E. N., Bettoney, W. E., and Pierrard, J. M., "A Total Vehicle Emission Control System," Paper No. 13-73, Presented to the API Division of Refining Meeting, May 15, 1973, Philadelphia, Pennsylvania.

12. Cantwell, E. N., Jacobs, E. S., Kunz, Jr., W. G., and Liberi, V. E., "Control of Particulate Lead Emissions from Automobiles," Paper No. 720672, Presented at the Society of Automotive Engineers National Automobile Engineering Meeting, Detroit, Michigan, May 22-26, 1972.

13. Environmental Protection Agency, "Promulgation of EPA's Proposed Regulations Affecting the Use of Lead and Phosphorus Additives in Gasoline" and Appendix C, "Analysis of Lead Traps vs. Removing Lead From Gasoline," September 1972.

14. "The Feasibility and Costs of Using Lead Traps as an Alternative to Removing Lead from Gasoline," E. I. du Pont de Nemours & Company, PLMR-6-72, 11/30/72; also letter from E. N. Cantwell to J. Somers, EPA, June 20, 1973.

15. Musser, G. S. and Berstein, L. D., "Development of an Automotive Particulate Sampling Device Compatible with the CVS System," Presented to the American Ind. Chem. Eng. 72nd National Meeting, St. Louis, Missouri, May 22, 1972.

16. Ter Haar, G.L., Lenane, D. L., Hu, J. N., and Brandt, M.,
    "Composition, Size, and Control of Automotive Exhaust
    Particulates," Journal of Air Pollution Control Association, 22,
    39, 1972.

17. Gentel, J. D., Manary, O. J., and Valenta, J. C., "Characteriza-
    tion of Particulates and Other Non-Regulated Emissions from
    Mobile Sources and the Effects of Exhaust Emission Control
    Devices on These Emissions," Report by Dow Chemical Company
    to U. S. Environmental Protection Agency, Office of Air and Water
    Programs, Ann Arbor, Michigan, (APTD 1567 pp. 179-184),
    March 1973.

18. LaPointe, Clayton, "Factors Affecting Vehicle Fuel Economy,"
    Paper No. 730791, Presented to the Society of Automotive
    Engineers National Meeting, Milwaukee, Wisconsin, September
    10-13, 1973.

19. Misch, H. L., Statement Before the U. S. Senate Public Works
    Committee, Washington, D. C., November 5, 1973.

20. Misch, H. L., Statement to the U. S. House of Representatives
    Subcommittee on Public Health and Environment of the Interstate
    and Foreign Commerce Committee, Washington, D. C., December
    4, 1973.

21. Cole, E. N., Statement and Attachments to the U.S. Senate Public
    Works Committee, Washington, D. C., November 5, 1973.

22. Private Communication from Esso Research & Engineering, Linden,
    N. J. to E. I. du Pont de Nemours & Company, Petroleum Labora-
    tory, Wilmington, Delaware.

23. Austin, T.C. and Hellman, K. H., "Passenger Car Fuel Economy -
    Trends and Influencing Factors," Paper No. 730790, Presented to
    the Society of Automotive Engineers National Meeting, Milwaukee,
    Wisconsin, September 10-13, 1973.

24. Huebner, G. J., "General Factors Affecting Vehicle Fuel Consump-
    tion," Presented to the 1973 Society of Automotive Engineers
    National Automobile Engineering Meeting, Detroit, Michigan,
    May 15, 1973. Available in SAE Special Publication SP-383.

25. Terry, S. L., Statement to the U. S. House of Representatives Subcommittee on Public Health and Environment of the Interstate and Foreign Commerce Committee, Washington, D. C., December 4, 1973.

26. Train, R. E., Statement and Attachments presented to the U. S. House of Representatives Subcommittee on Public Health and Environment of the Interstate and Foreign Commerce Committee, Washington, D. C., December 3, 1973.

27. Cole, E. N., Statement to U. S. House of Representatives Subcommittee on Public Health and Environment of the Interstate and Foreign Commerce Committee, Washington, D. C., December 4, 1973.

28. Glass, W., Kim, D. S., and Kraus, B. J., "Synchrothermal Reactor System for Control of Automotive Exhaust Emissions," Paper No. 700147, Presented to the Society of Automotive Engineers Automotive Engineering Congress, Detroit, Michigan, January 12-16, 1970.

29. Lang, R. J., "A Well-Mixed Thermal Reactor System for Automotive Emission Control," Paper No. 710608, Presented to the Society of Automotive Engineers Mid-Year Meeting, Montreal, Quebec, Canada, June 7-11, 1971.

30. "Octane Number Requirement Survey" 1963 through 1973, Coordinating Research Council, Inc., 30 Rockefeller Plaza, New York, New York.

31. Morris, W. E., Rogers, J. D., and Poskitt, R. W., "1971 Cars and the New Gasolines," Paper No. 710624, Presented to the Society of Automotive Engineers Mid-Year Meeting, Montreal, Quebec, Canada, June 7-11, 1971.

32. Bigley, H. A. and Benson, J. D., "Octane Requirement Increase in 1971 Model Cars - With and Without Lead," Paper No. 730013, Presented to the Society of Automotive Engineers International Automotive Engineering Congress, Detroit, Michigan, January 8-12, 1973.

33. Rogers, J. D., "Estimated Octane Requirements of 1974 Cars," Petroleum Laboratory, E. I. du Pont de Nemours & Company, Wilmington, Delaware, November 15, 1973.

34. "An Economic Analysis of Proposed Schedules for Removal of
    Lead Additives From Gasoline, " Bonner and Moore Associates,
    Houston, Texas, July 12, 1971. Prepared for the Environmental
    Protection Agency, Document PB 201 133, U. S. Department of
    Commerce.

35. Corner, E. S. and Cunningham, A. R., "Value of High Octane
    Number Unleaded Gasoline, " Presented to the Division of
    Petroleum Chemistry, American Chemical Society, Los Angeles,
    California, March 28-April 2, 1971.

36. Wagner, T. O. and Russum, L. S., "Optimum Octane Number for
    Unleaded Gasoline, " Paper No. 730552, Presented to the Society
    of Automotive Engineers Automobile Engineering Meeting, Detroit,
    Michigan, May 14-18, 1973.

The Application of the High Speed Diesel Engine as a Light Duty Power Plant in Europe

C. J. HIND

Perkins Engine Co., Peterborough, PE15NA, England

The fact that the diesel engine has been considered and used as a saloon car power unit for some 40 years may come as a surprise to some people   The diesel engine has succeeded in getting a name as a smelly, noisy, and rather smoky power unit, But things have changed and the days when only an enthusiast or an eccentric would drive a diesel powered car are almost past. The design and combustion features of the modern diesel engine are showing to be more compatible with the strict legislative demands thrust upon us and more people are now looking for a vehicle with good reliability, long life and maximum fuel economy.

But when did it all begin and why? The early 1930's really saw the first production high speed diesel engines. Prior to that the only engines had been of the heavy bulky industrial, marine type with a maximum speed of around 1000/1200 RPM, quite unsuitable for vehicle applications.

However, the fuel economy of the Diesel Cycle combined with the low fuel cost at that time made progression into the commercial vehicle market a natural move and the rapid development of high speed engines from the mid-1920's to the mid-1930's enabled the commercial vehicle operator to achieve lower operating costs thereby helping this market to rapidly expand. With the advent of the Bosch fuel injection equipment in Germany and later when C.A. Vandervell took up the manufacture of Bosch equipment in England, real strides were taken in market development.

The high speed diesel engine, with rated speeds of 3000 RPM plus came to be developed for use in the light truck market by the replacement route.

The diesel engine manufacturer offered to replace an existing gasolene engine in an operator's vehicle.

This developed a whole new design philosophy because inter-changeability and first cost being the only factors, the diesel engine had to fit into the space vacated by the gasolene engine, and provide a similar speed and torque range.

It was realised early in the development of the high speed diesel engine that cylinder pressures and engine breathing were going to be prime reliability and performance barriers.

The adoption of an indirect chamber engine allowed the intake port to be concerned only with inducing as high a mass of air as possible, and the swirl properties required for efficient combustion were provided by the air movement into and out of the chamber. Many designs of chambers were evolved during this time, each with its own theory.

One of the earliest and most successful designs was the Benz, later Mercedes Benz, pre-chamber or pepper pot design. (Fig. 1). This type of chamber has certainly stood the test of time as it is still widely used today and in many sizes of engines. Another successful chamber is the well-known Ricardo Comet chamber, which is still very widely used today in its refined form. (Fig.2).

My own Company, Perkins Engines Company, was formed in 1932 specifically to manufacture high speed diesel engines for the lighter class of vehicle and today manufactures high speed diesel engines at the rate of over 350,000 per year worldwide. As previously mentioned, inter-changeability with the gasolene engine wherever possible was the primary aim.

Fig. 3 shows comparative acceleration data taken in 1933 from a road test of a 4.2 ton GVW truck when fitted with its original 3 litre, six cylinder gasolene engine, and a 2.19 litre four cylinder diesel engine which replaced it without transmission or axle change. The similarity between the two curves was very encouraging at the time, especially when the fuel consumption of 15 mpg for the gasolene engine and 25 mpg for the diesel was also considered. The rated speed of 3000 RPM was the same for both types of engine. The savings due to the substantially better fuel economy of the diesel engine were even more enhanced when one considers that gasolene in Great Britain in 1933 cost the equivalent of 17 cents per gallon, whereas diesel fuel cost only 5 cents per gallon. The main reason for the difference was because gasolene fuel tax was some eight times higher than that on diesel fuel. In France diesel oil cost about half the gasolene price, and in Germany there was an even greater differential of approximately 70%.

Fig. 4 shows a comparative set of running costs that were issued in 1933 by the Commercial Motor. The considerably lower fuel costs are an obvious point, but the lower maintenance costs, even though the diesel engine was a new type of power unit, shows the other virtue of the diesel engine. The diesel engined vehicle had a 20% lower maintenance cost than the gasolene engined vehicle. (1)

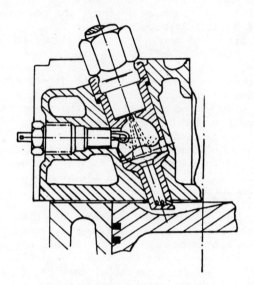

Figure 1.   Mercedes Benz chamber

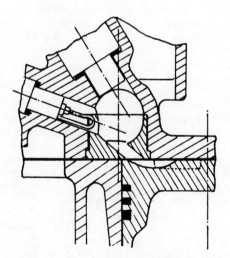

Figure 2.   Ricardo Comet chamber

This Utopia for the diesel engined vehicle could not last, and in Great Britain in 1934, our legislation penalised the diesel engine by a higher road tax. (Fig. 5).(2)

One novel fact that was put forward was that the increased motor taxes could lead to more deaths. The reasoning behind this statement being that more people would now go back to horse driven carts, and these beasts attracted flies which killed more people by infection than did the motor vehicle by road accidents.

Further pressure was applied to the diesel engine in 1935 when the British Government realised that there was a danger to its gasolene revenue, and so increased the tax on diesel fuel and made it equal to that on gasolene.

A number of statements made at the time make interesting reading such as the Minister's statement that "The oil engine can do as much work on 1 gallon of fuel as the petrol can do on 1-3/4 gallons". The increase of tax even pleased some people as it would "encourage the steam vehicle trade". Times don't change that much do they?

This increase of tax was a considerable blow to all concerned in the diesel market, and nearly bankrupted my company as they were committed to capacity expansion, but work continued as the better fuel economy of the diesel was still worthwhile, but it now became even more essential that the first cost should be maintained as low as possible.

The fuel injection equipment was, and still is, an expensive component in relation to the total engine first cost. But it soon became apparent that the reliability of the fuel injection equipment was considerably better than that of the electric ignition equipment fitted to the average gasolene engine. Consequently, the lower maintenance and down time costs were soon seen as a further bonus to the diesel engine vehicle operator.

The first diesel powered saloon cars. During the early 1930's, it became possible that the excellent fuel economy of the diesel engine would also prove attractive to the private motorist, and so the early 1930's saw parallel tests being run in passenger cars.

The first production diesel engined car was the Mercedes Benz "260D" which was powered by a four cylinder 2.6 litre engine which gave 45 HP at 3000 RPM. The car was normally fitted with a 2.3 litre gasolene engine.

The passenger car application was also being looked at in England by my company in the early 1930's with an eye to developing a market for Diesel conversion. In 1933, a 2.9 litre Perkins engine was installed in a gasolene production car as an option and a creditable running cost of 1/8 cent per mile was obtained.

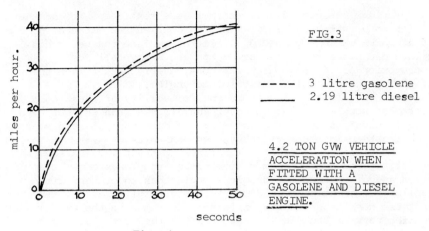

FIG.3

---- 3 litre gasolene
———— 2.19 litre diesel

4.2 TON GVW VEHICLE
ACCELERATION WHEN
FITTED WITH A
GASOLENE AND DIESEL
ENGINE.

Fig. 4

	Gasolene Engined Vehicles				Diesel Engined Vehicles		
	2 ton	3 ton	4 ton	5 ton	3 ton	4 ton	5 ton
Fuel	1.33	1.80	2.10	2.63	0.38	0.44	0.55
Lubricants	0.06	0.07	0.09	0.09	0.12	0.16	0.16
Tyres	0.28	0.35	0.44	0.49	0.56	0.74	0.84
Maintenance	1.23	1.42	1.57	1.70	1.15	1.26	1.35
Depreciation	0.54	0.66	0.93	1.05	0.80	1.10	1.27
Total	3.44	4.33	5.13	5.96	3.01	3.70	4.17

RUNNING COSTS (OLD PENCE PER MILE) IN 1933 IN GREAT BRITAIN

Fig. 5

Weight Unladen	Gasolene Pneumatic Tyres		Diesel Pneumatic Tyres		Diesel Solid Tyres *
	1933	From 1.1.34	1933	From 1.1.34	From 1.1.34
UNDER 12 cwt	£10	£10	£10	£35	£46
12 cwt-1 ton	£15	£15	£15	£35	£46
1-1-1½ ton	£20	£20	£20	£35	£46
1½-2 ton	£25	£25	£25	£35	£46
2-2½ ton	£28	£30	£28	£35	£46

  * FOR GASOLENE ENGINED VEHICLES WITH SOLID TYRES THE
    ROAD TAX REMAINED UNCHANGED AT THE SAME RATE AS THE
    PNEUMATIC TYRE TAX.

CHANGES IN THE VEHICLE ROAD TAX IN GREAT BRITAIN IN 1934

Various capacity diesel engines were tested and one of the bigger conversions was a 3.8 litre Gardner engine rated at 83 BHP at 3200 RPM which replaced a 3.5 litre gasolene engine. This saloon car had a top speed of 83 mph and an overall fuel consumption of 44 mpg, which was considerably better than the 16-18 mpg achieved with the gasolene engine.

The excellent fuel economy and reliability of these cars attracted people who had to cover very long distances, but even greater benefits were to be seen by the operators of stop start vehicles such as small delivery vans and taxis.

The diesel engine has nominally a constant volumetric efficiency and compression ratio through the load range at a given speed, whereas the gasolene engine has to contend with falling values at part load due to the throttling of the air flow at these conditions. This difference is shown in the better part load economy of the diesel engine and so the stop start or part load applications show the diesel engine to considerable advantage.

Fig. 6 shows how the specific fuel consumption curves of the same capacity engine when tested in diesel and gasolene forms diverge at the part load condition. This feature when transferred to actual road running results show that the light load running gives approximately three times the fuel saving seen at the high load factor running. (Fig.7).

Resulting from this characteristic and due to the high cost of fuel (gasolene or diesel) the rapid increase in the use of the diesel engine for taxi applications was most spectacular in Great Britain.

The first Diesel taxi was only registered in 1953 and yet within 2 years the number of new registrations had overtaken that of the gasolene engined taxis. (Fig.8).

Fig. 9 shows the population of diesel passenger cars in the German market with an impressive figure of 0.5 million diesel cars being used in 1972. A further 45,000 taxis were registered in Germany during 1973, and 80% of these were diesel powered. The position in France since 1963 is also shown, and although the actual numbers involved are much smaller, the trend shown from 1969 - 1972 is parallel to the German experience. Diesel fuel in France is still considerably cheaper than gasolene.

In Great Britain where fuel costs are the same and have been for a long time, we do not see the same usage of diesel engined cars in private use and consequently the majority of these vehicles are used as taxis. A typical difference in the fuel consumption for a London taxi cab type of duty would be 20 mpg for the gasolene engined taxi and the 35 mpg for its diesel engined equivalent.

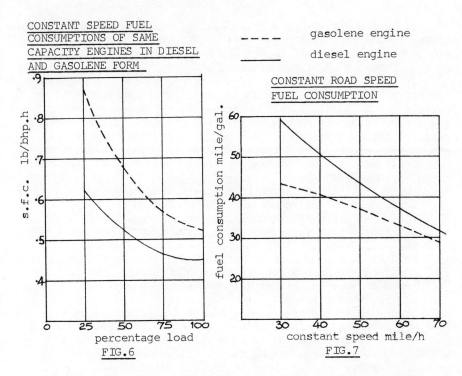

CONSTANT SPEED FUEL
CONSUMPTIONS OF SAME
CAPACITY ENGINES IN DIESEL
AND GASOLENE FORM

- - - - gasolene engine

——— diesel engine

CONSTANT ROAD SPEED
FUEL CONSUMPTION

percentage load

FIG.6

constant speed mile/h

FIG.7

NUMBER OF NEW REGISTRATIONS OF TAXIS IN GREAT BRITAIN.

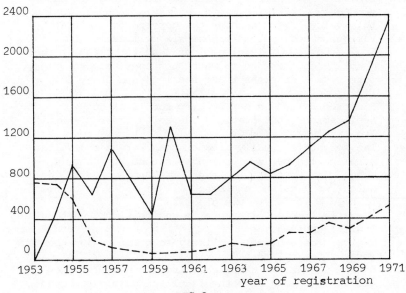

year of registration

FIG.8

If we look at the production rate of the diesel engined car in Europe over the last 15 years (Fig.10) we see that there has been a positive increasing rate which supports my opening comment on Diesel Engine use in motor cars.

It is of interest to note how the cost of fuel in European countries has affected the sale of Diesel Engined cars for private use (Fig.11). in all countries except Germany where there was a cost advantage between diesel and gasolene, diesel engined cars are quite common in private use. In England they are not and in Germany the cost differential has only just disappeared.

Until recently the limitation to the use of diesel engines could be summarised as follows:

1. Power   -   This is dependent on speed (rev/min) and brake mean effective pressure (B.M.E.P.).

2. Speed   -   For the size of engine considered, the limiting factor is usually mean piston speed. Problems arise if diesel engines are operated for sustained periods at piston speeds over 2500 ft/min. Some gasolene engines operate at up to 3500 ft/min. Fig.12 shows the permissible stroke dimension for various maximum engine speeds.

3. Stroke to Bore Ratio   -   For indirect injection diesels, a stroke/bore ratio of between 1.0 and 0.85 is possible. This therefore sets a limit on cylinder capacity for a given rev/min. and piston speed. Fig.13 shows the permissible maximum speed of various capacity six cylinder engines.

4. B.M.E.P.   -   Normally aspirated diesel engines should produce 90 lbf/in^2 b.m.e.p. at maximum speed. Using this value, Fig. 14 shows the horsepower limit at various rated speeds for the six cylinder engine.

5. Engine bulk   -   Diesel engines tend to be longer than gasolene engines due to water passages between bores, more robust crankshaft and bearings and heavy duty timing drive.
   Siamesed cylinders may be used for light duty applications, but problems due to cylinder distortion are likely.
   The height is usually greater than for an equivalent gasolene engine, due to longer stroke and thicker head. Carburettors, however, frequently add to the height of gasolene engines. Oil pans tend to be deep to hold a larger volume of oil.
   There is little difference in engine width, particularly in-line engines.
   The bulk of a diesel is likely to be up to 50% greater for a given cylinder capacity.

6. Engine weight   -   Where cast iron is used for the blocks and heads of both diesel and gasolene engines, the diesels are usually heavier. This can amount to

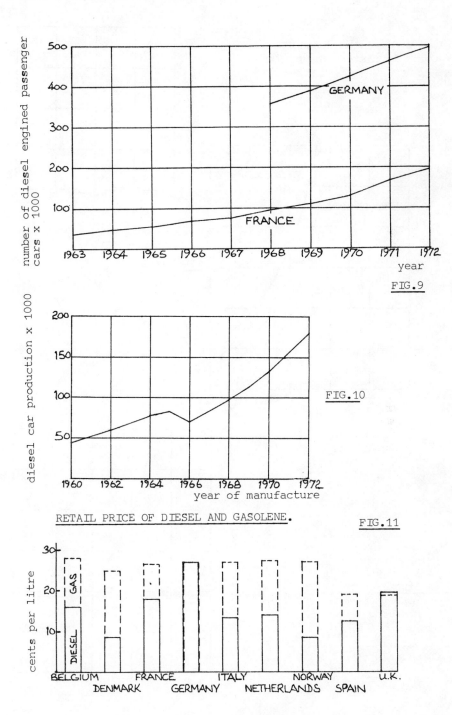

FIG.9

FIG.10

RETAIL PRICE OF DIESEL AND GASOLENE.   FIG.11

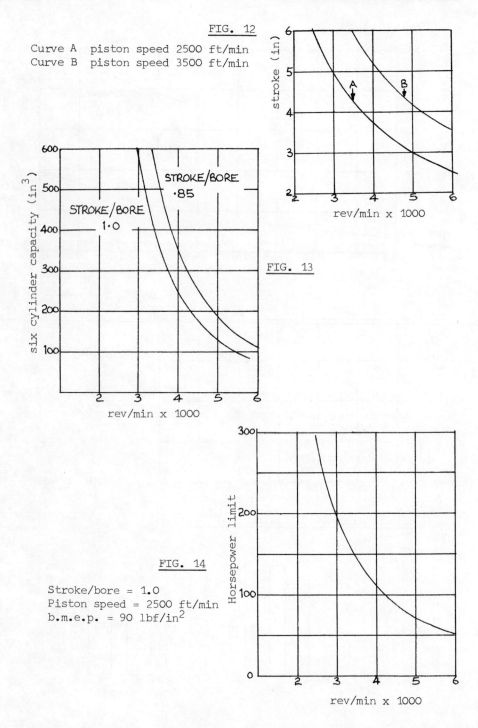

FIG. 12

Curve A piston speed 2500 ft/min
Curve B piston speed 3500 ft/min

FIG. 13

FIG. 14

Stroke/bore = 1.0
Piston speed = 2500 ft/min
b.m.e.p. = 90 lbf/in^2

100% more for equal power, normally aspirated.

Fig. 15 summarises these disadvantages over the gasolene engine.

FIG. 15		
	Diesel	Gasolene
Cylinder Block Length	27.6 in	26.1 in
Engine Length	36.6 in	31.0 in
Height above Crankshaft	18.4 in	18.3 in
Overall Height	29.0 in	26.9 in
Overall Width	23.2 in	22.8 in
Engine Weight	708 lb	555 lb
Net Horsepower	105 at 3600 rev/min	110 at 4000 rev/min

COMPARISON OF 6 CYL. DIESEL AND GASOLENE ENGINES.

So what is the Future? — particularly in the light of legislative and fuel resource pressures.

Two approaches are possible — one being to turbocharge as a short term possibility — and the other is to apply new design techniques for longer term adoption.

Fig. 16 shows a comparison of test bed performance between a 4 cylinder, 108 cubic inch turbocharged diesel engine and a 4 cylinder, 104 cubic inch gasolene engine.

Superiority of diesel clearly seen!

Fig. 17 shows the comparative performance of both engines when tested in a standard European passenger car, and Fig. 18 the comparative road fuel consumption results at constant road speeds.

Recent developments at Perkins have led us to believe that we can now design a diesel engine which avoids the limitations outlined in Fig. 15.

Emission control work has enabled us to lower maximum cylinder pressures and rates of pressure rise, to very near gasolene engine figures. This would enable a diesel motor car engine to be designed — which would have all the fuel economy characteristics of the current engine — but would be little heavier than the equivalent gasolene, no more noisy — and only marginally more expensive.

The late 1970's will see this engine!

Literature Cited

1.   "The Commercial Motor" (1933) Vol 58 page 268

2.   "The Commercial Motor" (1933) Vol 58 page 380

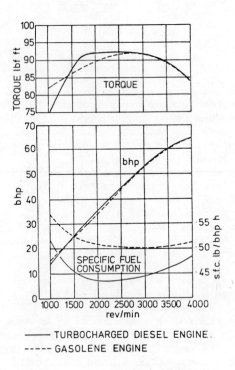

TURBOCHARGED DIESEL ENGINE.
GASOLENE ENGINE

Figure 16

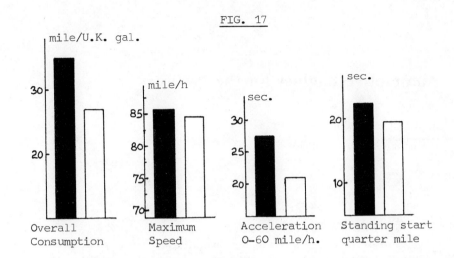

FIG. 17

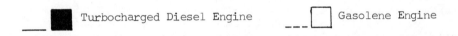

Turbocharged Diesel Engine   Gasolene Engine

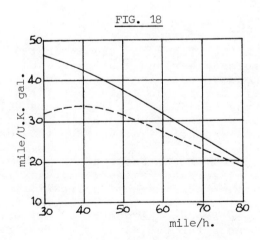

FIG. 18

CONSTANT SPEED FUEL CONSUMPTION

9

Automotive Engines for the 1980's

R. W. RICHARDSON

Eaton Corp., Southfield, Mich. 48075

Abstract

There are five major contenders to replace or supplement today's piston engine. Changing social requirements and new technological developments will lead
to major changes in automotive power plants. This
paper makes projections through the 1980's of the market penetration of Wankel, Stirling, Turbine, stratified charge and diesel engines for passenger car, heavy
duty and small engine applications. These engines are
compared on the basis of ten major selection parameters.
Major factors affecting the rate of commercialization
of new engines are reviewed including social, political
and economic forces of change and historical perspective. Major inputs came from more than 60 worldwide
in-depth interviews.

I   Introduction

Never in the history of the automotive engine have
there been so many serious contenders and never with so
great a chance of replacing or supplementing the piston
engine. While the piston engine has for many decades
served its hundreds of millions of users well and is
continuing to serve them well, its noise, exhaust emissions and more recently its fuel appetite have come
under attack.

The purposes of the basic study upon which this
report is based were to assess various new engine
types, determine their market applicability and likely
commercialization through the 1980's and provide broad
overall perspective on the future of automotive engines.

Major inputs for the study were obtained from over
60 in-depth interviews worldwide. These included car
and truck manufacturers; heavy duty and small engine

producers; developers of new engines; materials, parts, fuels and lubricants suppliers; machine tool builders; government agencies; trade associations; independent research institutes and consultants. These inputs were combined with business, technical and historical analyses and an evaluation of the social, political and economic forces that cause change.

Primary emphasis was placed on the Wankel engine and on those factors which will have the greatest bearing on its (degree and rate of) commercialization. Priority was placed on passenger car application followed closely by heavy duty markets with a relatively modest effort in the small engine area.

II Engines and Applications

The engines that are now used and those which warrant and/or are receiving serious attention for three broad areas of application--small engines, passenger cars and heavy duty--are listed in Table I.

Candidate Alternate Powerplants by Market

	Small Engine	Passenger Car	Heavy Duty
NOW	2-Cycle- 4-Cycle	4-Cycle	Diesel
FUTURE	Rotary Turbine	Rotary Turbine Stirling Stratified Charge Diesel	Turbine Stirling

Table I

The passenger car area has the most contenders. Electric and steam vehicles have not been included as serious contenders for high volume applications as a result of Eaton's previous in-depth investigations (not published).

Like the piston engine, these new engines come in a wide variety of broad configurations and sub-types. The type or configuration of each engine believed best suited for each application has been used in the comparative evaluation for that application.

These different powerplants are all in various stages of development (Figure 1) which has a bearing on the ability to assess accurately the various parameters important to engine selection: factors such as fuel

consumption can be accurately determined but production
cost or durability are little more than guesses at an
early stage of development. Engines at an early state
of development are also much more susceptible to rapid
improvement than mature engines.

III   Descriptions & Status of Engine Types

Rotary (Wankel). The Wankel engine is a four-cycle
spark ignition internal combustion engine differing
from present engines primarily in mechanical design.
It uses a "rotating" (epitrochoidal) motion rather than
a reciprocating motion. It uses ports rather than
valves for controlling the intake of the fuel-air mixture and the exhaust of the combusted charge. In this
respect it is similar to a two-cycle engine.
    The Wankel has been under development since its
invention in the early 1950's. NSU (West-Germany) introduced a single rotor powered car in 1974 and a two
rotor powered car in 1966. Neither engine has been
built in significant volume. Toyo-Kogyo (Japan) introduced a two rotor engine in their Mazda cars in the
late 1960's and are now producing about 20,000 models
per month. 50,000 of these were sold in the U.S. in
1972 and nearly 100,000 in 1973.
    Snowmobiles with a Fichtel and Sachs single rotor
Wankel have been sold for the past four seasons. Similar engines are being used in power lawn mowers. Outboard Marine Corp. is producing Wankel snowmobiles and
has outboard versions under development. A number of
major firms throughout the world own a Wankel license
and several--most notably General Motors--are aggres-
sively pursuing its development. GM announced plans to
produce 100,000 Wankel powered Vegas for the '75 model
year. Introduction has been delayed until early 1975
while improvements are made in fuel consumption.

    Turbine. The gas turbine engine is a continuous
flow, continuous internal combustion, high speed engine
utilizing aerodynamic compression and expansion rather
than positive displacement. The turbine uses no valves
or ports. The engine requires a number of parts made
from high temperature alloys.
    The gas turbine has been under development since
the 1930's. It has found ready acceptance in aircraft
and is now nearly universally used except below 500 HP.
The auto industry has been working on turbine power for
nearly 25 years. Exotic show cars have been produced
from time to time. About 10 years ago, Chrysler produced 75 special turbine powered cars for field testing.

Most of the industry's effort has since then been directed at truck and industrial applications. Both GM and
Ford have produced pilot quantities of an industrial
engine of about 300 HP. Ford has recently closed down
their pilot operation to await a major product redesign
aimed at the late 1970's. The big three all are reported to have substantial passenger car turbine development programs. The Environmental Protection Agency
is funding part of Chrysler's program and also passenger
car turbine development by several aircraft engine producers.

   Stirling. The Stirling engine is an external continuous combustion engine utilizing positive displacement piston compression and expansion. It utilizes a
sealed high-pressure working fluid (hydrogen or helium)
and operates at relatively low speed. High temperature
alloys are required for the combustor-to-working-fluid
heat exchanger (heater head).
   The Stirling was invented in 1816 and saw service
as a pumping engine in mines during the 19th century.
These engines used air at low pressure as the working
fluid. The modern Stirling engine dates from the late
1930's based on work by N. V. Philips in the Nether-
lands. In recent years, considerable progress has been
made in refining the Stirling engine.
   Philips has licensed other developers. During the
1960's GM was Philip's major licensee, accumulating
more than 25,000 hours of engine operating experience
in their development program. GM allowed their license
to lapse in 1970, however. More recently (1972) Ford
Motor Co. was licensed by Philips. They are jointly
working on passenger car prototypes. Philips has also
licensed United Stirling in Sweden and MAN in Germany.
United Stirling has been a major contributor to recent
progress and is also working on passenger car applications. Both Philips and United Stirling have made recent prototype bus installations.

   Stratified Charge. The stratified charge or hybrid
engine is a variant of conventional engines combining
features of both gasoline and diesel engines. It differs from conventional gasoline engines in that the
fuel-air-mixture is deliberately stratified so as to
produce a rich mixture at the spark plug while maintaining an efficient and cleaner burning overall lean mixture and minimizing or avoiding the need for throttling
the intake air.
   Stratified charge development dates back at least
to the work done by Ricardo in England during World

War I. Since that time, many inventors and developers
have worked with various concepts. Substantial work
has been done in Russia and the U.S. over the past 15
years. More recently, the Japanese, especially Honda,
have made major contributions to the state of the art.
The Honda CVCC engine meets the original 1975 emission
standards without hang-on controls. Ford and Texaco
have done substantial work on concepts quite different
from Honda's.

   Diesel. Diesel engines are quite similar to gasoline engines but use fuel injection directly into the
cylinder rather than a carburetor. They have no ignition system as such, relying on very high compression
to cause the mixture to self ignite.
   Diesel engines have been widely used in heavy duty
applications for decades. They are also used to limited extent on passenger cars mainly in Europe. Mercedes has long produced a low performance diesel car.
More recently, Peugeot and Opel have been building
diesel cars. Many are used for taxis. Austin (BLMC)
also builds diesel taxis. The engine has not been
seriously considered for passenger cars in the U.S.
It is now receiving considerable attention due to the
energy crisis.

IV Engine Selection Parameters

   Table II lists the more significant engine selection parameters.

Engine Selection Parameters

Traditional	New
Cost	Emissions*
Durability - Life	Noise*
Weight	
Size	
Smoothness	
Flexibility	
Maintenance	
Fuel Consumption*	

*Social Requirements

Table II

They are all self-explanatory except for flexibility
which means performance flexibility or torque-speed
characteristics as they relate to transmission requirements and driveability. The parameters listed on the
left are traditional ones. On the right, two new parameters are listed which are primarily social requirements. Fuel consumption has also been labeled a social
requirement because of the energy crisis--it has long
been an economic or logistic requirement.

Some idea of the changing relative importance of
the social requirements can be gained from Table III.
They have been rated on a 0-10 scale to provide helpful
perspective. Only a slight lessening in absolute importance of emissions (assuming no major air pollution
disasters) is expected--some increase in the importance
of noise and a tremendous increase in the importance of
fuel consumption--becoming half again as important as
emissions before the end of this decade. Gasoline
prices have increased very substantially over the past
two years and further increases are likely. Unless
these higher prices have a greater than expected impact
on demand, some form of rationing is likely to be
needed during the next few years.

Today, of course, as the chart shows, emissions
are more important. There is wide speculation on Congress revising the very stringent and now delayed
1975-76 standards to achieve a better balance between
society's needs for acceptable cost and fuel consumption as well as emissions.

The significance of the emission levels which are
ultimately selected is their great bearing on both absolute and relative cost and fuel consumption of different engines. As emissions are reduced, both cost
and fuel consumption tend to increase for all different
types. They are likely to increase, at markedly different rates, however, for different engines. For
example, a low-cost conventional piston engine may
require a very-expensive precious-metal dual-catalyst
to meet a tight standard while a somewhat more complex
and costly stratified charge engine, such as the Honda,
may require no extra emission controls.

V Selection Parameters & Comparison of Engine Types

Three areas of application have been considered:
passenger cars, heavy duty and small engines. Obviously, the priority of selection parameters differs
somewhat in each of these three areas.

A. Passenger Cars. Taking passenger car applica-

tions first, the selection parameters fall into the order of relative importance shown in Table IV with flexibility, smoothness and emissions leading the list, and
maintenance, fuel consumption and durability on the
bottom. This ranking is for 1973 values. The arrows
on the left show both noise and fuel consumption rising
to expected 1980 positions. This order of importance
of parameters is for the broad passenger car market.
Obviously, there may be segments of this market which
would have somewhat different orders of importance.
The five new engine types competing for future automotive use are compared with the 4-cycle, spark ignition
piston engine on each of these parameters. Each engine
was rated better (+), worse (-), or equal (·0) to the
present gasoline engine.

   Wankel. (1) Flexibility. Wankel engines tend to
have lower torque at low speeds and a higher speed for
their torque peak than reciprocating engines. This
means that the rotary engine has reduced performance
flexibility requiring more shifting or more sophisticated transmissions for equivalent performance.
   (2) Smoothness. A single rotor Wankel can be
equal to or better than many 4-cylinder engines, while
a two rotor Wankel is smoother than 4- or 6-cylinder
engines. Most interest has been generated for two
rotor Wankel engines with only little interest in three
and four rotor engines and very little interest in single rotor engines.
   (3) Emissions. Untreated, the Wankel is a rather
dirty engine with emissions of hydrocarbons as much as
five times higher, carbon monoxide up to three times
higher while oxides of nitrogen are up to 75% less.
Derating a conventional engine to the same level of
efficiency would be expected to result in similar emission levels. Conversely as the Wankel seals are improved, oxides of nitrogen will tend to increase and
hydrocarbons decrease. The Wankel has fewer exhaust
ports and, because it is less efficient, operates with
a higher exhaust temperature which makes thermal reactors more applicable.
   (4) Cost. Although the Wankel uses fewer parts
and is lighter, even when built in high volume, it costs
more than a piston engine. It will likely be more expensive for some time to come; however, with substantial
product and manufacturing development effort, it could
ultimately become cheaper to produce. (The cost is
discussed in more depth in Section VIII.)
   (5) Noise. The elimination of mechanical moving
parts, such as the valve gear, should reduce noise but

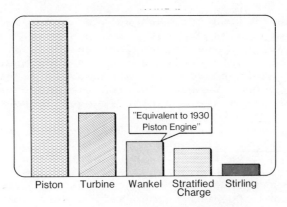

Figure 1. Relative state of engine development

Table III. Relative Importance of Social Requirements

	Aug 72	Feb 73	Mar 74	1975	1980	1985
Emissions	7	7	6	6	6	6
Noise	3	3	3	3	4	4
Energy Resources	1	3	7	10	7	6

Table IV. Relative Importance of Selection Parameter

Passenger Cars	Compared with 4-Cycle Spark Ignition Piston Engine				
	Wankel	*Turbine	Stirling	Stratified Charge	Diesel
Flexibility	–	+	+	0	–
Smoothness	+	++	++	0	–
Emissions	0	+	++	+	+
Cost	–	–	–	?	–
Noise	0	+	++	0	–
Weight	+	+	0	–	–
Size	+	0	–	–	–
Maintenance	0	+	+	0	+
Fuel Consumption	–	–	++	+	++
Durability	–	?	+	0	+

Advantage (+) or Disadvantage (−) *Two-Shaft Regenerative 1900 F Turbine Inlet Temperature

data on the limited car (and snowmobile) models available show approximately equal noise levels.

(6) Weight. Wankel engines weigh less, especially
when only the basic engine structure is compared. When
completely equipped with all accessories needed for
operation, there is a smaller relative advantage as
these accessories are essentially equal or occasionally
heavier.

Wankel weight savings of 50% are often claimed.
Realistic comparison shows much less. Comparison of
the lightest experimental Wankel known with the lightest production piston engines indicates 12-16% weight
savings on a pounds per horsepower basis. (These engines are not of equal horsepower.) Karl Ludvigsen, in
a recent article in "Road Test" mazagine, indicates an
average weight savings of 11% comparing several pairs
of engines of equal performance. Significant reductions
in average weight and size of piston engines are possible should this become a high design priority. (Weight
is further discussed in Section VI.)

(7) Size. Wankel engines also have a size advantage usually somewhat greater than their weight advantage. Most of the comments on weight (above) also
apply here. Comparing the lightest and most compact
engines shows size advantages in the range of 35-45%
based on a "box" volume (max. length X max. height X
max. width). Karl Ludvigsen's analysis indicated a 30%
average advantage. (The significance of weight and size
savings as it relates to packaging in an automobile is
discussed in Section VII.)

(8) Maintenance. The maintenance requirements of
the Wankel are expected to be reasonably comparable to
the piston engine. The Wankel has fewer but more com-
plicated and expensive parts. It uses similar types of
fuel, air cleaning, ignition, cooling and exhaust systems. Wankels currently use a more complex emission
control system requiring more maintenance.

The Wankel can provide some savings through longer
oil change periods. A recent survey indicates tune-up
costs at dealers are approximately equal for Mazda and
V8 engines. Fours and sixes cost less to tune up of
course.

(9) Fuel consumption. Fuel consumption includes
both quantity and quality of fuel. Wankel engines have
substantially higher fuel consumption: 30-40% higher
(or 25-30% fewer miles per gallon) than piston engines.
At very low emission levels, this difference will probably be reduced but not eliminated. Improved seals will
also help reduce, but not eliminate, this fuel consumption penalty as the combustion chamber appears inher-

ently less favorable. In contrast to its higher use of
fuel, the Wankel has a requirement for lower octane
quality (low to mid-80's for Mazdas). The octane requirement will probably increase as seals are improved
and as the engine is scaled up to larger displacement
per rotor.

(10) Durability. The durability of Wankel engines
was initially very poor. The Mazdas have substantially
better durability but are not yet up to typical U.S.
standards. Compatible trochoid surface materials and
treatment, together with seal materials having very long
life, have been developed but apparently not with acceptable cost and sealing characteristics.

Based on this somewhat superficial comparison,
weighted for importance of parameters, the Wankel
appears to have little, if any, net advantage. As the
fuel consumption issue takes on more importance, the
Wankel's competitive position will be more tenuous. If
manufacturing cost breakthroughs are achieved, it may
still find a substantial niche.

Turbine. (1) Flexibility. Two-shaft turbines
have a very favorable torque curve having in effect a
built-in torque converter. Single shaft-engines which
have recently come under serious consideration have an
unfavorable torque curve.

(2) Smoothness. As a continuous fluid flow rotary
machine, the turbine is extremely smooth.

(3) Emissions. Turbine combustors can be built
which have very low emissions, especially of hydrocarbons and carbon monoxide. There is some question as to
whether they can meet the pending 1976 NOx standard of
0.4 grams/mile.

(4) Cost. Turbine engines require the use of substantial amounts of expensive, difficult to fabricate
superalloys and an expensive regenerator. Potentially
possible, but requiring a great development effort, is
a simpler turbine operating at higher pressure ratios
and temperatures using lower cost ceramic materials.
Such an engine could ultimately be cheaper than the piston engine. Single-shaft engines cost significantly
less than two-shaft, but require more sophisticated and
costly transmissions.

(5) Noise. Despite the image of turbine powered
aircraft, the turbine engine is relatively easy to
silence.

(6) Weight. Turbine engines are substantially
lighter than piston engines.

(7) Size. The basic turbine is also substantially
smaller than the piston engine but the addition of a

regenerator results in no net size advantage.

(8) Maintenance. The turbine is basically a simple machine and, with freedom from vibration, should have lower maintenance.

(9) Fuel consumption. The turbine has higher fuel consumption, especially at light loads typical of much automobile operation. At very tight emission standards, the fuel consumption increase of piston engines could result in the disadvantage of the turbine being eliminated. Development of materials allowing operation at higher temperature would help make the turbine competitive on fuel consumption. The turbine is capable of operating on a wide range of fuels, but specific designs require a limited range.

(10) Durability. Aircraft turbine engines have demonstrated much greater durability than piston engines. There is considerable doubt, however, whether this will be true for cars due to operation with very frequent wide fluctuations in load and operation with dirtier air. The addition of the regenerator required for reasonable part load fuel consumption may also reduce durability.

Overall, the turbine appears to have a significant potential net advantage and apparently warrants additional development effort.

Stirling. (1) Smoothness. Stirling engines are also extremely smooth engines effectively completely balanced and have very minor cyclical variations in torque.

(2) Flexibility. The Stirling engine has a favorable torque curve providing substantial torque increase as speed falls.

(3) Emission. The Stirling, based on bench tests, appears to have the lowest emissions of all known engines, well within 1976 requirements--achievable with little penalty in fuel consumption or cost.

(4) Cost. The Stirling appears to have a cost disadvantage due to the requirement for high temperature alloys in the heater head and to control problems. Recent developments indicate these control problems are not as formidable as once believed. At very tight emission standards, the piston engine could conceivably increase sufficiently in cost to make the Stirling competitive or possibly give it an advantage.

(5) Noise. The Stirling engine has a very low noise level and is the quietest of any of the serious contenders.

(6) Weight. Recently developed double acting Stirling engines appear competitive in weight.

(7) Size. Stirling engines are somewhat larger
than piston engines but studies show they can be installed with all accessories in engine compartments of
both sub-compact and full size cars.
      (8) Maintenance. Like the turbine, the Stirling
should have relatively low maintenance requirements.
      (9) Fuel consumption. The Stirling has a fuel
consumption potential lower than any other contender
and will operate on the broadest range of fuels.
Achievement of the very low fuel consumption may not be
possible with a practical size radiator. Compromise
would still leave the engine with lower fuel consumption
than any engine but the diesel.
      (10) Durability. Developmental Stirling engines
have shown extremely high durability--perhaps due to
over-design. Some compromise to help reduce cost may
be in order.
      On balance, the Stirling engine appears potentially
the most attractive powerplant over the long range.

      Stratified charge. (1) Flexibility. The strati-
fied charge engine can equal the flexibility of the piston engine, although it is difficult to achieve. It
may not be achievable on all types of stratified charge
engines.
      (2) Smoothness. The stratified charge engine as
a modification of the gasoline engine should have approximately equal smoothness.
      (3) Emissions. Stratified charge engines have
shown potential for low emissions. Honda has readily
met the original 1975 standards with both small and
large cars and has come close to 1976 standards with
small cars. Other types of stratified charge engines,
such as the Ford Proco have shown potential for relatively low emissions (but not yet as good as Honda).
      (4) Cost. As there will probably be some loss in
maximum power and some increase in complexity (pre-chamber 3rd valve or fuel injection), some increase in cost
over present (1973) engines is likely. However, compared to dual catalysts, it would likely be substantially cheaper.
      (5) Noise. Noise should be equivalent.
      (6) Weight. Due to some probable loss in maximum
power, relative weight would slightly increase.
      (7) Size. The same applies for size as for
weight.
      (8) Maintenance. Stratified charge engines should
require slightly more maintenance than uncontrolled en-
gines--somewhat less maintenance than engines with dual
catalyst.

(9) Fuel consumption. Probably a slight advantage
in fuel consumption will be realized by practical stratified charge engines although development will be required. Burning of overall lean mixtures and reduced
pumping losses both save fuel. Stratified charge engines can use a broader range of fuels.

(10) Durability. Based almost exclusively on current piston engine technology, the stratified charge
engine's durability should be similar.

The stratified charge engine, on balance, is not
as attractive as the Stirling, but because it is based
largely on existing parts, it could be commercialized
relatively rapidly with only slight disruption to the
industry. It is only marginally attractive compared
with current engines. It is quite attractive, however,
compared to conventional engines with dual catalyst.

Diesel. (1) Flexibility. Diesel engines generally operate over a narrower speed range and require
more gear ratios and shifting.

(2) Smoothness. The diesel is less smooth than
the gasoline engine due to its combusiton characteristics.

(3) Emissions. Diesel engines have very low hydrocarbon and carbon monoxide emissions and can have
fairly low oxides of nitrogen emissions but probably
not low enough to meet 1976 requirements. Diesel engines, however, also tend to produce objectionable
smoke and odor.

(4) Cost. Because of the direct fuel injection
system (15-25% of engine cost) and because of the requirement for a more rugged structure, diesel engines
are substantially more costly.

(5) Noise. Diesel engines are generally much
noisier.

(6) Weight. Unless the diesel were turbo-charged
to a high boost-pressure and run at high speed, it
would be substantially heavier.

(7) Size. The same applies for size as for
weight.

(8) Maintenance. Diesels have proven to have low
maintenance requirements primarily due to their heavy
rugged design.

(9) Fuel consumption. The diesel is a very efficient engine and has a substantial fuel consumption advantage especially at light loads characteristic of
passenger cars.

(10) Durability. The diesel has also been proven
to be a very durable engine also due to its rugged design.

On balance, the diesel does not appear to be an attractive alternate for passenger cars.

Stratified charge Wankel. Much has been made recently of the potential for a stratified charge Wankel. Operating on the same thermodynamic principle as the piston engine, it is possible to produce stratified charge Wankel engines. Because of the gross mechanical design differences in the two engines, it is usually not possible to have exactly equivalent stratification approaches. The development of a stratified charge Wankel represents a greater technical challenge than development of the conventional Wankel and therefore appears to be further away. Stratification should have relatively the same advantages and disadvantages as it does for the piston engine although due to the specific designs evolved, their relative costs might be quite different.

B. Heavy Duty. For heavy duty applications, the parameters have been reordered (Table V) with fuel consumption, maintenance and durability moving from least important to most important, flexibility and smoothness move from most important to near least important. The basic engine for comparison is the diesel engine.

The Wankel engine considered here is substantially different from the passenger car Wankel. The high compression ratio required for diesel compression ignition results in very unfavorable geometry in a Wankel. Engines of this type have been built but performed very unsatisfactorily. This problem can be overcome by compounding two stages of lower compression. Rolls Royce is the apparent leader with this approach. The two-stage Wankel diesel is in a much earlier state of development and the problems to be overcome are greater. It has the disadvantage of adding complexity but still results in a very compact engine. The Wankel requires only one fuel injector for each two-stage unit. This version will probably be built primarily with two-stage units. The two fuel injectors required compare with 6 or 8 on a piston type diesel engine. Fuel injectors must operate at twice the frequency at the same engine speed and as the speed of the Wankel is higher than the piston engine, the maximum frequency of injection is much higher. Injection equipment to operate at these frequencies has not yet been developed.

While this type of Wankel has advantages compared with piston diesels, they are in the least important parameters. The Wankel, therefore, looks unattractive for heavy duty application.

Both the turbine and Stirling not only appear far more attractive than the Wankel, but also offer advantages over the diesel. The turbine's high fuel consumption may, however, prevent it from achieving substantial acceptance. The Wankel diesel might have potential for medium duty applications where fuel consumption, maintenance and durability are of lesser importance.

C. Small Engines. Small engines are used in a wide variety of applications such as chain saws, lawnmowers, generator sets, pumps, and recreational vehicles and low-power industrial vehicles. The requirements of these engines can vary significantly. For purposes of analysis here, the parameters have been placed in order of importance for consumer product applications (Table VI). Both two-and four-cycle piston engines are used for these products. The two-cycle engine is predominant at the higher power levels of most recreational vehicles. It has, therefore, been chosen as the base engine.

Wankel engines for this application would be primarily single rotor and would use charge cooling of the rotor instead of oil cooling and would use an oil-fuel mix similar to many two-cycle engines.

There appears to be no way the Wankel can ever be cost competitive with single cylinder piston engines. Although Wankels are offered in snowmobiles, outboards, lawn mowers and for model airplanes, they are sold to a very limited market at very substantial premiums. The Wankel has a better chance against multi-cylinder engines (above 15 HP). As the products using small engines become subject to more stringent noise and emission regulation, the cost, size and weight of the piston engine will rise faster than for the Wankel, perhaps making it competitive. The Wankel has advantages in all the other parameters except durability which is relatively unimportant. The Wankel, therefore, looks promising for recreational vehicles especially those now using multi-cylinder engines.

VI Passenger Car Engine Packaging

Size, weight, and configuration of engines are significant factors in engine installation and vehicle layout. The piston engine is being used in a wide variety of vehicle configurations today--front engine, rear engine or mid-engine; longitudinal or transverse mounting; and with front or rear drive. Size and weight of engines have been relatively unimportant (ranked 6th and 7th of ten parameters) in the past but are expected

Table V.   Relative Importance of Selection Parameter

Heavy Duty Applications	Compared With Diesel Engines		
	Wankel	Turbine*	Stirling
Fuel Consumption	−	−	+
Maintenance	−	+	+
Durability	−	?	+
Emissions	−	0	+
Noise	0	+	++
Cost	?	−	−
Smoothness	+	++	++
Flexibility	0	+	+
Weight	+	+	+
Size	+	0	0

Advantage (+) Or Disadvantage (−)   *Two-Shaft regenerative 1900 F Turbine Inlet Temperature

Table VI.   Relative Importance of Selection Parameter

Small Engine Applications	Compared With 2-Cycle Spark Ignition Engine
	Wankel
Cost	−
Weight	−
Size	0
Flexibility	+
Smoothness	+
Noise	+
Emissions	+
Maintenance	+
Durability	−
Fuel Consumption	+

Advantage (+) or Disadvantage (−)

to become increasingly important due to the trend toward
smaller cars and the increase in space requirements and
weight of emission control and safety equipment. The
"energy crisis" will further increase the importance of
size and weight.

Major factors behind projections of a rapid Wankel
revolution are claims of savings of 50% or more in engine size and weight. These savings are projected to
be further amplified through the redesign of passenger
cars permitting a net reduction in car length of 30
inches without sacrificing passenger space due to the
smaller engine. This reduction in length, together with
the lighter engine, has been projected to result in a
total weight reduction of 1,000 lbs., together with con-
sequent cost savings.

Wankel proponents claim that these goals can be
reached by placing the engine across the front of the
car using front wheel drive--a so called transverse engine. Indeed, the claimed weight and size advantages
for the Wankel are considered so compelling to both user
and manufacturer as to cause a rapid changeover to the
Wankel engine for economic and competitive reasons.

Extensive analysis of size and weight of the Wankel
engine, however, reduced this claimed advantage somewhat
as discussed previously (Section V). Weight savings of
10-20% and volume savings of 30-40% seem reasonable.
These still represent substantial improvements, especially in volume. However, these volume savings are not
so readily converted to major reductions in vehicle
size. The volume comparisons are based on the box
created by multiplying the maximum width, height and
length. Most piston engines do not fill the box very
completely. Within the box, there is considerable
space for accessories, frame rails, running gear, sus-
pension components, or emission controls.

In contrast, the Wankel engine is indeed very compact but much more nearly fills its box. The Wankel
engine is both somewhat lower and shorter than a typical 4 cylinder piston engine but not narrower.

In the Toronado front wheel drive car, the transmission and differential are placed alongside the
engine adding only slightly to width and height of the
engine "box". This is the type of arrangement which
would also be used with a transverse installation.
Packaging a transmission and differential with a Wankel
engine would add significantly to the width and/or
height of the installed Wankel engine box.

Considerable length saving can be achieved with
present engines (Figure 2). The 1972 LTD shown here
has more than 18 inches of unutilized or poorly utilized

space ahead of the engine. Some cars have even more
waste space. Since the time the stylist hid the radiator, more than 40 years ago, unused or poorly utilized
space in front of the engine has been the rule. The
'47 Ford had about 14". A number of Ford's competitors,
shortly after this time, switched from in-line 8-cylin-
der engines to V8 engines, saving considerably on engine
length, but their cars did not get shorter or the passenger compartments larger--in fact the reverse was
true. Waste space is present in intermediate and compact cars and also in many foreign cars.

Perhaps the best example of a very effective packaging job is the family of cars BMC introduced in 1959
(Figure 3). The 52 cubic inch displacement (C.I.D.)
Mini was following shortly with the 67 C.I.D. MG and
then the 110 C.I.D. Austin America, all with transverse
engine front wheel drive. More recently, a 136 C.I.D.
transverse 6-cylinder model has been produced. Within
the same width, which is 12" narrower than full sized
U.S. cars, it would be possible to put a 272 C.I.D. V-12
engine--a V8 of the same length (car width) and having
same B/S ratio could be as large as 615 C.I.D.--nearly
25% larger than the largest car engine in production.
These very compact vehicles did not, however, have such
compelling advantages as to cause a massive switch to
this concept.

These transverse engine cars are no longer imported
into the U.S., having been replaced by a conventional
front engine rear drive model. Nevertheless, many
others have adopted the concept.

The Toyo Kogyo Mazda RX2 with the Wankel engine in
a conventional rear drive arrangement is shown at the
top of Figure 4. Recently introduced in the U.S. is
the Honda CIVIC model with a 72 C.I.D. engine. This
car with a 91 C.I.D. CVCC stratified charge engine
(which meets the '75 emission standards) has recently
been introduced in Japan.

There appears to be no way the Mazda engine could
be fit, regardless of position, in the space available
in the Honda, Fiat, Peugeot or BLMC cars without increasing the vehicle length. There appears to be no
substantial packaging advantage in conventional configuration either, at least against 4-cylinder engines.
Significant length savings could be achieved compared
with a 6-cylinder engine, however. One Wankel disadvantage with front engine rear drive also shown in Figure 4
is the higher driveshaft requiring a higher tunnel
through the car. This is due to the centerline of the
Wankel rotor shaft being significantly higher than a
piston engine crankshaft.

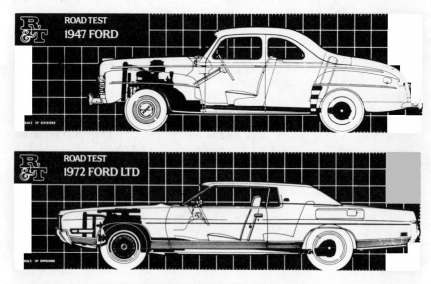

Figure 2.   1947 and 1952 Ford

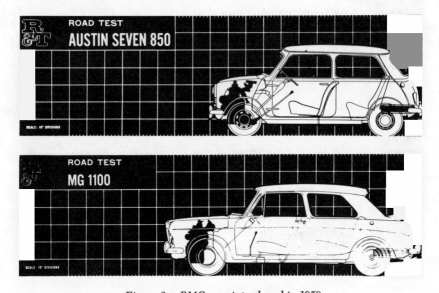

Figure 3.   BMC cars introduced in 1959

These silhouettes indicate that major reductions in the size of piston powered cars are achievable. The volume savings of the Wankel are simply not capable of being converted to substantial additional savings in size, weight or cost. Furthermore some of the realizable weight saving of the Wankel is due to the substantial use of aluminum.

VII   Design Flexibility

Much has been made of the design flexibility of the Wankel--allowing any number of rotors to be "stacked" allowing common tooling to provide for a very wide range of power needs. There is very little interest in single rotor engines for passenger cars so the minimum engine became a two-rotor engine. Three and four-rotor engines have been built, but there is a cost penalty of a two-piece rotor shaft and coupling or a split gear and housing to allow more than two rotors to be assembled. To assemble more than four rotors requires a second split. Wankel rotors may be "stacked" in-line only, without going to multiple shafts and additional gearing or chain drives. In contrast, cylinders may be arranged in-line, vee, or opposed configuration and in-line engines may have their cylinders mounted horizontally, vertically or slanted as needed to fit a variety of applications.

The turbine engine has even less design flexibility than the Wankel. The Stirling engine can have design flexibility equaling the piston engine but economics may reduce this flexibility potential somewhat.

This design and application flexibility is very important to the relatively low volume producer. Using a large number of common parts, a wide range of power needs, from 70 to 1400 HP, can be covered with one cylinder size by Detroit Diesel by varying the number of cylinders (Figure 5). Cummins covers a range of 200 to 800 HP using one cylinder size with only 6- and 12-cylinder engines. This flexibility allows engines to be designed specifically for some very low volume applications, but still at a reasonable cost.

VIII   Cost

Cost is the greatest unknown factor and will probably have the most influence upon the applicability and rate of commercialization of new engines. Cost is a function of weight, number of parts and technological density. The latter concept comes from Mr. Yamamoto of Toyo Kogyo. Technological density includes materials,

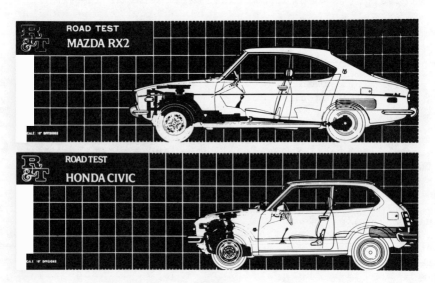

Figure 4.   Mazda RX2 and Honda Civic

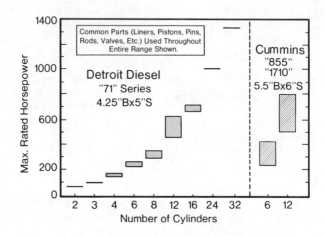

Figure 5.   Design flexibility of the piston engine

quantity and quality of machined surfaces, tolerances,
surface treatments, and design complexity. Most cost
projections are superficial, being based on the first
two terms only.

This is a widely used preliminary estimating approach and is valid when the third term is relatively
equal, but in the case of at least Wankels and turbines,
which do weigh less and have fewer parts, it is not.
The Wankel uses 42 or more sealing elements per rotor
(some configurations use up to 58 per rotor). In contrast, the piston engine uses only five per cylinder.
Including the valve gear parts and the seals, the piston
engine uses only 19 per cylinder. The piston engine has
the advantage that most are duplications of simple parts
produced in high volume--eight of each valve gear part
(16 each on an 8-cylinder engine).

In contrast to the piston rings, which are self-
retained in place, Wankel seals are held in place only
by gravity or friction with the seal springs trying to
eject them. This suggests some difficult assembly problems not readily lending themselves to economical auto-
mation.

The only examples of production Wankels, whether in
cars, snowmobiles or model airplanes, currently sell for
high premiums. Mazda sells for a $600 premium here and
more than a $300 premium in Japan where the thermal
reactor emission controls have not been used. Mazda's
cost penalty is not known but believed to be substantial. While Mazda's volume is low (156,000 Wankels/
yr - 1972) compared to the highest volume U.S. engine,
it is not low compared with their piston engine production: in 1972 it accounted for 40% of passenger car
production. It is actually comparable in volume to
many U.S. engines including Chevrolet, AMC or Ford 6's,
AMC 8's and several other 8's.

Since the Wankel's introduction in 1967, Toyo Kogyo
sales have more than doubled but their profits have
steadily declined. Their average retail premium worldwide was about $400 and with normal discounts, about
$250 per car was received by Toyo Kogyo. It appears
most of this premium is eaten up in higher costs.

U.S. snowmobiles with Wankel engines sell for a
premium of from $160-$350. A small model airplane engine sells for nearly four times the cost of an equivalent two-cycle piston model airplane engine.

One of the areas of high cost is materials for and
machining of the trochoid surface. While a great many
material and treatment combinations have been tried, the
most successful have been aluminum housings with either
chrome by the Doehler-Jarvis transplant process, Elnisil

(or tungsten carbide). Toyo Kogyo and NSU are believed
to be using equipment capable of grinding four to five
housings per hour (2 to 2-1/2 engines/hr). Recent reports indicate this may have doubled using diamond
grinding wheels. Several U.S. machine tool builders
have recently introduced proto-type grinders capable of
finishing 20-25 housings per hour (10-12 engines/hr).
     Typically, piston engines are produced at 100 or
more/hr. It seems likely that another generation of
machine tools will be required before high volume production would be economically practical. Even if a cast
iron housing could be used without treatment and produced at substantially higher rate, there is doubt
whether the Wankel could be produced competitively. So
far, cast iron does not look feasible.

IX Capital Investment

     Closely related to cost are capital investment re-
quirements. The U.S. auto industry and its suppliers
are estimated to have invested, at replacement cost,
over $50 billion in machinery and equipment (not includ-
ing plant and land). At least 15 to 20%, or $8-10 bil-
lion, is estimated to be in the engine production area.
The auto manufacturer's greatest annual investment in
machinery and equipment has been about $2 billion.
Only a portion of these investments, of course, are in
machine tools.
     The machine tool industry has an annual capacity of
about $2 billion but probably not more than 20-25% can
be devoted to the auto industry. Ralph Cross, president
of the Cross Co., told the Environmental Protection
Agency (1973) that a changeover to a completely new
engine would take 12.3 years, based on present capacity
of the transfer (automated manufacturing) machine industry. This could probably be improved upon resulting in
a 10 year conversion if warranted. There may be other
limiting factors. The lead time to equip the industry
with trochoid grinders (even for slow rate grinding) is
apparently not one of them.

X Other Commercialization Considerations

     Further insight into the possible rate of commer-
cialization may be gained from a look at history of
other major new automotive innovations. Figure 6 shows
the history of automatic transmissions, disc brakes and
air conditioning. The Automatic Transmission took 20,
Air Conditioning--17, and Disc Brakes--9 years from
successful introduction to a 50% market penetration.

All had been marketed unsucccessfully in the U.S. many
years earlier.

Disc brakes, for example, were used in production
on Chrysler cars as early as 1950. They also had been
very widely used in Europe. In addition, they report-
edly cost less to produce in high volume than drum
brakes--a strong incentive for rapid change.

Power steering and power brakes took 14 and 17
years respectively to reach 50% market penetration.
These are examples of the most successful products.
Many lesser successes have not reached 10% and some have
approached 10% only after 15 to 25 years.

Many others didn't make it at all--air ride lasted
less than two years despite nearly a 3% initial penetration. Fuel injection lasted nine years but had a maximum penetration of less than a tenth of one percent.

Probably the best analogy that can be made is with
the change to modern short stroke overhead valve engines
(Figure 7). This first shows how the engine market mix
has changed since the mid-30's. Superimposed (whiskered
area) is a curve showing the transition to the modern
engine. This change started in 1948 and was not complete until 1966--18 years later--evolutionary, not
revolutionary. In 1948 the industry was due for a
change: the last previous significant new engine was
introduced 17 years earlier and some engines were approaching 30 years of age. Production tooling was
largely obsolete and worn out. Furthermore, substantial
R&D had taken place on engines and considerably higher
octane fuels had become available, both largely in response to wartime aircraft needs.

The steep increase in the mid-50's of both modern
engines and eight cylinder engines seems to parallel
the horsepower race. Pent-up demand and consumer savings resulting from two wars helped fuel this growth.
Had a long term smooth growth curve for V-8 engines
taken place, the transition curve would probably have
been a typical smooth S curve with the mid-point about
1955. This early growth represents the demonstrated
capacity of the industry for major substantial engine
change and tends to suggest this conversion to modern
OHV engines could probably have been completed in ten to
twelve years.

Today the situation is quite different--last year
Ford brought on stream a new engine plant to produce
the Pinto engine in Lima, Ohio. Ford is also building
a new engine plant in Brazil. Two other engines are
less than four years old and almost all engines in production have been introduced or retooled within the
past decade. In addition, there is a much wider range

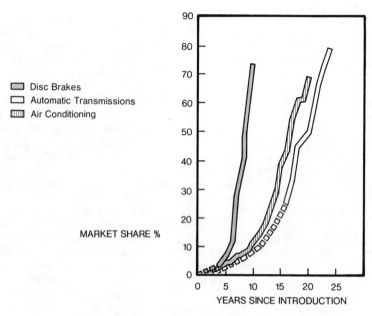

Figure 6.   *Commercialization history of major automotive products*

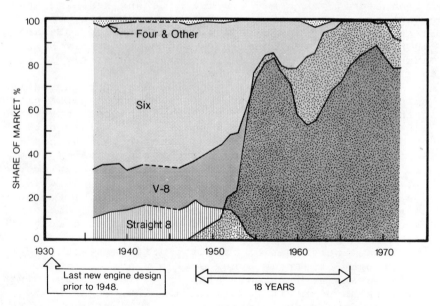

Figure 7.   *Commercialization history of modern OHV engines*

and greater number of both engine and car models in
production. A complete transition could, therefore,
take longer today.

The heavy duty engine producers also have a number
of relatively new engines in production or under development for near term introduction, and new plans are
under construction to produce diesel and heavy duty
gasoline engines.

If indeed substantial cost savings are achieved--
more than 15% (and some project up to 30 or 40%)--then
there would be incentive to obsolete unamortized tooling
make these large investments, and move rather quickly to
the Wankel. With today's demands for corporate social
responsibility, the major auto companies could only move
rapidly to the Wankel if its serious fuel consumption
penalty were substantially reduced. The limiting factor
then would likely be the machine tool industry.

On the other hand, if there is little or no cost
saving, commercialization will be at a rather slow pace,
and at a cost penalty, will be quite limited. 55,000
Mazdas at a $600 premium is one thing. There may even
be a market of as many as 500,000 U.S. built sporty
novelty cars at a few hundred dollars premium--but certainly not a market for 10 million.

XI   Market Share Projections

Passenger Car. Taking all the previously discussed
factors into account and applying both optimistic and
pessimistic assumptions (within a reasonable range)
yields the projected range of probable market share for
Wankels shown shaded in Figure 8. The maximum probable
is about 13% by 1980 and 23% by 1985. The minimum probable rises to 3% in the late 1970's, gradually fading
away in the mid-1980's. Developments of the past year
indicate that the likely use will be toward the lower
side of this band. The maximum possible curve is based
on the assumptions of greatly increased R&D, major
manufacturing developments and the capacity of both the
machine tool and auto industries to finance and produce
the necessary production equipment. In addition, some
major incentive for this rapid change which is not now
apparent would be required.

Other new engine types are in the picture and must
be considered. How long will there be to amortize the
investment before it is obsoleted by one of the other
advanced powerplants. Similar projections show that
at the maximum possible penetration the turbine reaches
50% only 6 years after the Wankel--long before the production investment would normally be amortized.

Figure 9 is a composite. It repeats the Wankel projections of Figure 8 shown shaded in the lower part of the
chart. The light area at the lower right shows the
penetration of the turbine and Stirling engines, beginning no earlier than 1980 and rising to a maximum of 8%
in 1985. The minimum could readily be zero.

The balance of the market, the reciprocating piston engine, is obtained by subtracting the sum of turbine and Wankel minimum and maximum penetrations from
100%. It would have a market share of at least 69% and
could conceivably take the whole market in 1985. The
maximum piston engine market share in 1980 is 97% due
to the forecast minimum Wankel penetration. Beginning
with the 1975 models, catalytic exhaust treatment will
be widely employed to enable these piston engines to
meet emission standards as shown by the steeply rising
"hump" in the middle of Figure 9.

The catalyst curve shows an early decline as the
stratified charge engine comes into use. The stratified charge engine may indeed prove sufficiently attractive to not only take over this whole reciprocating
engine segment, at least 69% of the total, but to even
recapture the small segment lost to the Wankel in the
mid-and late-1970's.

Heavy Duty. Projections for the turbine engine's
penetration of the heavy duty market are shown in
Figure 10. Ford recently announced they were closing
down their pilot production line after building 200
engines to await an improved new design late in the
1970's.

The Stirling engine could not be introduced before
the early 1980's and would likely not exceed the upper
limit of the probable turbine curve at least until the
very late 1980's.

The Wankel, if it comes, will be later and slower.
Heavy Duty engine manufacturers are more cautious and
move slower than passenger car manufacturers. (Most
have new diesel engines under development and some new
plants under construction to produce diesels.)

Small Engines. Cost will prevent the Wankel from
competing effectively against single cylinder engines
except in very limited premium markets--less than a 5%
penetration would be expected. The Wankel looks much
better compared with multi-cylinder engines used in
recreational vehicles (Figure 11). Recreational vehicles include snowmobiles, motorcycles, ATVs, outboard
motors, etc. The Wankel could replace more than half
of these piston engines by the mid-eighties in this

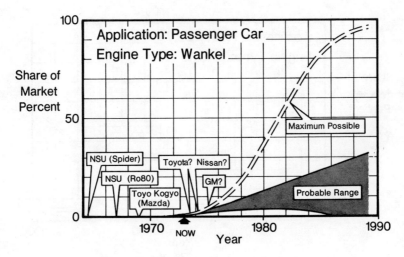

Figure 8.   Future engine use projections

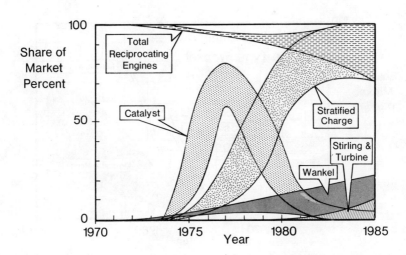

Figure 9.   Range of expected market penetration

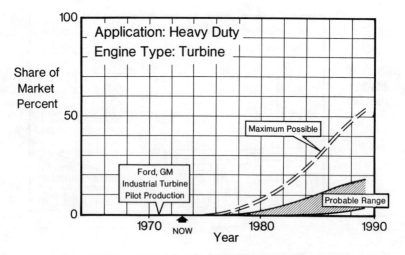

Figure 10. Future engine use projections

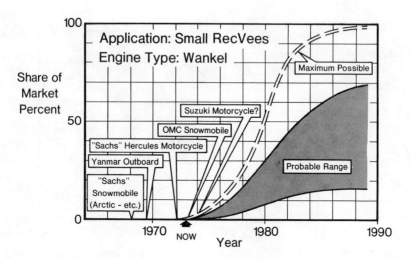

Figure 11. Future engine use projections

application. Despite the greater optimism, there is
also greater uncertainty reflected in a wider band in
this application area. After 7 years of offering a
single rotor Wankel in their snowmobiles, Artic Cat
(the 2nd largest snowmobile producer) has discontinued
their Wankel model.

Geographic Consideration. These projections generally represent both the U.S. domestic and worldwide
picture. In the case of the Wankel, Japan will move
faster and Europe slower. Europe is ahead on the Stirling with Japan last. The U.S. is ahead on passenger
car turbines. The U.S. is also ahead on heavy duty
turbines with Europe next and Japan last. Japan is
ahead on stratified charge engines and Europe apparently last.

XII Summary and Conclusions

Reciprocating piston engines will remain dominant
well into the 1980's. The Wankel engine will receive
increased use in passenger cars, possibly approaching
a 25% penetration by the mid-1980's, but probably much
less.

Wankels will be more significant for small engines,
especially compared with multi-cylinder engines for
recreational vehicles, perhaps reaching 50% of this market segment, but they will not significantly penetrate
the small single cylinder engine market.

Wankels will not be significant in heavy-duty engine applications. Turbine engines have greater long
range potential in both cars and heavy duty applications and the Stirling engine probably even greater
potential.

In the near term, the stratified charge engine
looks potentially very attractive offering required low
emission performance without costly hang-on controls
and representing relatively minor tooling changes.

Vehicle and engine manufacturers continue to approach change with caution and will follow conservative
introduction and commercialization strategies. Economics will continue to be the dominant influencing factor,
but social requirements, especially fuel consumption,
will become more significant in influencing change to
different engines.

The overall conclusion, therefore, is that there
still is considerable uncertainty as to the choice and
rate of commercialization of specific engines, but no
revolutions are likely in the near future.

XIII   Acknowledgment

     This paper is based on a recently conducted major
technological planning study.   In addition to the au-
thor, the research team included L. F. Jenkins of
Eaton's Valve Division and R. P. Horan and R. L. Martin
of Eaton's Technological Planning Department.
     Scale side view drawings showing engine and drive
trains of various automobiles have been reproduced
from "Road & Track" magazine by permission of Bond/
Parkhurst Publications, a unit of CBS Publications.

INDEX

A

Acceleration with a gasoline and diesel
  engine, vehicle 160, 163
Air
  combustion rates, hydrocarbon– 79
  –fuel equivalence ratio 66
    carbon monoxide emissions and en-
      gine power as functions of 44, 49
    hydrocarbon emissions as a function
      of 43, 45
    on nitric oxide formation rate,
      influence of 80, 81
    nitrogen oxide emissions as a
      function of 44, 45
  monitoring program (CAMP),
    continuous 102
  pollution control, attainment of 101, 104
  quality standards 102
    attainment of carbon monoxide 112, 115
    for hydrocarbons, carbon monoxide,
      nitrogen dioxide, and photo-
      chemical oxidants, EPA 100
    vehicle carbon monoxide emission
      rate to meet 110, 111
  quality studies 112
  quality trend analysis 103
Alternate automotive emission control
  systems 99, 113, 135
Alternate powerplants by market,
  candidate 173
Analysis, air quality trend 103
Area traffic volumes in metropolitan
  centers 110, 111
Aromatic fuel manufacture 16
Automobiles, gasoline demand
  projections for 28, 31, 32, 33, 34
Automotive
  diesel engine emissions 94
  emission control systems,
    alternative 99, 113
  engines for the 1980's 172
  products, commercialization history
    of major 194, 196
  trends and emissions regulations 19

B

Back-end volatility 62, 63, 65
Bituminous coal hydrogenation 14, 15
BMC cars 189, 190

C

C$_5$–400°F reformate 71, 72
C$_6$ hydrocarbons, shape selective
  cracking of 73
C$_6$–350 Kuwait naphtha 73
C$_{26}$ saturated hydrocarbons, viscosity of 2, 4

California interim standards, 1975 140
CAMP (continuous air monitoring
  program) 102
Capital investment 194
Car (see also Automobile and Vehicle)
  BMC 189, 190
  design and equipment trends 21, 22
  diesel powered saloon 162
  engine packaging, passenger 186
  engines, passenger 177, 197
  power unit, saloon 159
  production, diesel 166, 167
  sales mixes, new 30, 31, 32, 34
  trends, passenger 21, 22, 30
  with Du Pont thermal emission control
    systems 142
Carbon as a source of energy, unaltered   6
Carbon dioxide emissions 66
Carbon monoxide
  air quality standard 112, 115
  concentration 103, 106
  emissions 61, 84, 91, 93, 95, 102
    flux density 109, 110
    as a function of air–fuel equivalence
      ratio 44, 49
    with gaseous fuels 53
    rate, vehicle 110, 111
    standards, vehicle 107, 108
  EPA air quality standards for 100
  EPA automotive emission standards for 100
  response to mid-range and back-end
    volatility 63, 65
  trends in six cities 107, 108
  values in Chicago, maximum 105, 106
Carburetion system, modifications to 63
Catalyst life 70, 76
Catalyst, zeolite 71
Catalytic
  emission control systems
    effect of proposed 137, 138
    on fuel economy, effects of
      137, 138, 141, 145, 146, 151, 152
    and non-catalytic systems 145
  reactor 69, 70
  unit 71, 72
Chamber
  configurations, combustion 80
  conventional and divided combustion 91, 92
  engine, divided 91, 92
  engine, indirect 160
  engine, single-cylinder divided
    combustion 93
  fuel injection, timed combustion 82
  Mercedes Benz 160, 161
  Ricardo Comet 160, 161
  staged combustion engine, divided- 90
  stratified charge engines, open- 82

Charge engine
open-chamber stratified 82
prechamber 86, 87
stratified 85
stratified 93, 175, 183
Charge stratification 85
Charge Wankel engine, stratified 185
Chemical fuels 1
Chevrolet
Du Pont muffler lead traps
on 1970 ....127, 128
Impala, Hondo CVCC conversion
of 350-CID 89, 90
standard and modified high
compression ratio 125, 126
with TECS III and TECS IV 123, 124
thermal reactor for a
350 CID 116, 119, 120, 121
Chicago, maximum carbon monoxide
values in 105, 106
Chicago traffic activity 103, 104
Cities, carbon monoxide trends in six 107, 108
Civic, Honda 88
Coal
energy recovery of hydrocarbons from 7, 11
hydrogenation 9
bituminous 14, 15
properties of products from 12, 13
liquid fuels from 16
weight recovery of hydrocarbons from 7, 11
Coke, energy recovery of hydrocarbons
from 6, 8
Coke, pure hydrogen from 6
Coke, weight recovery of hydrocarbons
from 6, 8
Combustion
chamber
configurations 80
conventional and divided 91, 92
fuel injection, timed 82
single-cylinder divided 93
engine, divided-chamber staged 90, 93
engines, low emissions 78
engines, modified 79
gases, engine 79, 81
heat of 1, 2, 3, 4
modification 79
processes, continuous 94
processes, timed staging of 80
rates, hydrocarbon–air 79
system (CVCC), Honda compound
vortex-controlled 86
system (PROCO), Ford-
programmed 82, 83
system (TCCS), Texaco-controlled ... 82, 83
Commercialization considerations, other 194
Commercialization history of major
automotive products 194, 196
Commercialization of modern OHV
engines 195, 196
Compression ratio 23
Chevrolet, standard and modified
high 125, 126
increase 33
reduction 113, 136, 138

Configurations, combustion chamber 80
Consumption, fuel (see Fuel
consumption)
Continuous air monitoring program
(CAMP), EPA's 102
Continuous combustion processes 94
Conversion of 350-CID Chevrolet
Impala, Honda CVCC 89, 90
Conversion to gaseous fuel 51
Conversion level, octane rating vs. 73, 75
Conversion of liquid hydrocarbons to
gaseous fuel 69
Conversions, Ford PROCO 85
Conversions, jeeps with stratified engine 84
Converter, pre-engine 69
Cost
in 1933 in Great Britain, running 160, 163
of conversion to gaseous fuel 51
as a function of octane number,
gasoline 12
of gaseous fuel 52
of new engines 191
Cracking of C_6 hydrocarbons, shape
selective 73
CVCC conversion of 350-CID Chevrolet
Impala 88, 89, 90
CVCC engine, Honda 86, 87
CVCC (Honda compound vortex-con-
trolled combustion system) 86
Cylinder diesel and gasoline engines, six 169
Cylinder divided combustion chamber
engine, single- 93

D

Density, carbon monoxide emission
flux 109, 110
Design, engine 16
Design flexibility of the piston
engine 191, 192
Design of TECS III and TECS IV 116
Diesel
car production 166, 167
engine 93, 176, 184
constant speed fuel con-
sumptions of 164, 165
emissions, automotive 94
fuel economy of 94, 162
high speed 159
six cylinder 169, 171
test bed performance of
turbocharged 169, 170
use of 166
vehicle acceleration with 160, 163
plant, gasoline– 16
powered saloon cars 162
retail price of 166, 167
Divided chamber engine 91, 92
Divided chamber staged combustion
engine 90, 93
Divided combustion chamber engine,
single-cylinder 93
Divided combustion chamber NO_x
emissions 91, 92
Dual fuel capability 48

Du Pont
  data 142, 143
  muffler lead trap 126, 131, 133
    on 1970 Chevrolet 127
    on 258 CID, 1971–72 Hornets 131
  thermal emission control systems 142
  total emission control system,
    (TECS) 114, 115

E

Efficiency of fuel synthesis processes,
  thermal 14
Efficiency improvement 33, 34
Efficiency, vehicle 32
EGR emission control systems, thermal
  reactor– 144, 145
Emissions
  automotive diesel engine 94
  carbon dioxide 66
  carbon monoxide ..61, 80, 84, 91, 93, 95, 102
  combustion engines, low 78
  control systems
    alternative automotive 99, 113, 135
    catalytic 141
    and thermal 145, 146, 151, 152
    Du Pont thermal 142
    lead tolerant 126
    proposed catalytic 137, 138
    (TECS), Du Pont total 114, 115
    thermal 143, 146
    thermal reactor–ECG 144, 145
  control with TECS III and IV 120
  controls
    exhaust 21
    on fuel economy,
      effect of 39, 40, 135, 152
    on fuel economy of 1968–1974
      model cars, effect of 136, 138
    potential effects of 25
  factors for gasoline and gas powered
    vehicles 53, 54
  flux density, carbon monoxide 109, 110
  as functions of air–fuel equivalence
    ratio 43, 44, 45, 49
  with gaseous fuels 53, 57
  Honda CVCC 88, 89, 90
  hydrocarbon 61, 80, 84, 91, 93, 95
  improvements with gaseous fuels 55, 56
  from jeeps with stratified engine
    conversions 84
  lead 129, 130, 131, 134
  levels of Ford PROCO conversions 85
  levels of TECS-equipped Pinto 120, 122
  limits 29
  NO$_x$ 79, 80, 84, 89, 91, 92, 93, 95
  with only engine modifications 125
  pollutant 55, 56
  rate
    allowable carbon monoxide 109
    lead 132, 133
    standard, vehicle carbon
      monoxide 110, 111
    vehicle carbon monoxide 105, 106
  reduction in hydrocarbon 43
  regulations on gasoline demand,
    impact of 19

Emissions (continued)
  response to mid-range and back-end
    volatility, hydrocarbon 62, 65
  standards
    air quality and 102
    federal exhaust 79
    on fuel economy, effect of 25, 26
    on gasoline demand projections,
      effect of 28, 31
    NO$_x$ 113
    predicting 105
    vehicle carbon monoxide 107, 108
  tests 93
  volatility characteristics and exhaust 62
Emitted lead particles from TECS Pinto   130
Energy
  calculations 6, 7
  consumption, patterns of 19, 20
  content of test fuels 68
  loss for manufacturing liquid fuels
    from coal 16
  recovery from bituminous coal
    hydrogenation 14, 15
  recovery of hydrocarbons from coke 6, 8
  recovery of hydrocarbons from shale
    and coal 7, 11
  resources, influence of fuel
    composition on 1
  unaltered carbon as a source of 6
Engine
  combustion gases 79, 81
  conversions, jeeps with stratified 84
  converter, pre- 69
  design 16, 191, 192
  development, state of 179
  emissions, automotive diesel 94
  emissions, single-cylinder divided
    combustion chamber 93
  modifications, emissions with only 125
  packaging, passenger car 186
  power as a function of air–fuel
    equivalence ratio 44, 49
  selection parameters 176
  types 174
  use projections 197, 198, 199, 200
Engines
  for the 1980's, automotive 172
  cost of new 191
  diesel (see Diesel engine) 93, 176, 184
  divided-chamber staged combustion 90
  gas turbine, stirling cycle, and
    rankine cycle 94
  gasoline (see Gasoline engines)
  heavy duty 185, 198
  high speed diesel 159
  Honda CVCC 86, 87
  indirect chamber 160
  low emissions combustion 78
  modified combustion 79
  OHV 195, 196
  open-chamber stratified charge 82
  passenger car 177, 197
    prechamber stratified charge 85
    reciprocating spark ignition 82
    rotary (Wankel) 174, 178
    small 186, 198

Engines (continued)
    stirling 175, 182
    stratified charge 93, 175, 183
      Wankel 185
    torch ignition 86
    turbine 174, 181
    turbocharged 169, 170
EPA air quality standards 100
EPA automotive emission standards 100
EPA's continuing air monitoring
    program (CAMP) 102
Equipment trends, new car 21, 22
ERG (exhaust gas recirculation) 114
Esso research and engineering data 141, 143
Exhaust emissions
    controls 21
    levels of TECS-equipped Pinto 120, 122
    standards, federal 79
    volatility characteristics and 62
Exhaust gas recirculation
    (EGR) 114, 120, 121
Exhaust manifold thermal reactor 114

F

Federal exhaust emissions standards 79
Flexibility, design 191
Flux density, carbon monoxide
    emission 109, 110
Ford, 1947 and 1952 ....188, 190
Ford PROCO conversions 85
Ford-programed combustion system
    (PROCO) 82, 83
Fossil fuels, hydrogen content of 2, 5
Fuel
    capability dual 48
    composition and total energy resources 1
    consumption calculation 25
    consumption, patterns of 19, 20
    consumptions of diesel and gasoline
      engines, constant speed 164, 165, 171
    economy
      from 1967 to 1974, change in 136
      of 1968–74 model cars 136, 138
      at 55°F ambient 66, 67
      assessment of 152
      of Chevrolets with TECS III and
        TECS IV 123, 124
      comparisons 135
      for current production vehicles 85
      of diesel engines 94, 162
      of divided chamber staged com-
        bustion engine 93
      with Du Pont thermal emission
        control systems 142
      effect of catalytic emission control
        system on 141
      effects of catalytic and thermal
        emission control systems on
        ....145, 146, 151, 152
      effect of emission controls on....39, 40, 135
      effect of emission standards 25, 26
      effect of proposed catalytic emis-
        sion control systems on 137, 138
      effect of thermal emission control
        systems on 143, 146
      of Honda engine 89

Fuel (continued)
    loss 27
    effect of octane requirement
      reduction on 149, 150
      on 91 octane unleaded gasoline .. 147
      of 1971 Pinto with and without
        TECS III 123, 124
      with standard and modified high
        compression ratio Chevrolet ..125, 126
      of thermal reactor–EGR emission
        control systems 144, 145
      vehicle 113
      to volatility, relationship of 68
    equivalence ratio, air– (see Air–fuel
      equivalence ratio)
    injection 82, 162
    manufacture, aromatic 16
    synthesis processes, of 14
    system, LPG 48, 49
    volatility 60, 62
Fuels
    chemical 1
    from coal, liquid 1, 16
    gaseous (see Gaseous fuels)
    hydrogen contents of fossil 2, 5
    from lower hydrogen content materials 6
    test 68, 71
Future gasoline demand, projecting....24, 30, 31
Future new car sales mixes,
    hypothetical 30, 31

G

Gas, natural 47
Gas-powered vehicles emissions
    factors for 53, 54
Gas-powered vehicles on pollutant
    emissions, effect of 55, 56
Gas recirculation (EGR),
    exhaust 114, 120, 121
Gas turbine, engines 94
Gaseous fuels
    carbon monoxide emissions with 53
    catalytic reactor converting liquid
      hydrocarbons to 69
    cost of 52
      conversion to 51
    hydrocarbon emissions with 53
    motor vehicle emission improvements
      with 55, 56
    NO_x emissions with 53
    power loss with 44
    retrofit modifications for 48
    safety of 51
    storage of 47
    sulfur dioxide emissions with 57
    supply of 50
Gaseous-fueled vehicles, performance of 50
Gases, engine combustion 79, 81
Gases, liquefied petroleum (LPG) 47
Gasoline
    consumption 19, 20
    cost as a function of octane number 12
    demand
      effect of lead in 33
      growth, moderating 29

Gasoline (continued)
   impact of automotive trends and
     emissions regulations on 19
   projections 24, 30, 31
   effect of efficiency improvement
     on 33, 34
   effect of emission standards on 28, 31
   effect of new car sales mix on ...32, 34
   –diesel plant 16
   engines
     constant speed fuel consumptions
       of 164, 165
     performance of 169, 171
     six cylinder 169
     test bed performance of 169, 170
   412 ton GVW vehicle acceleration
     with 160, 163
   fuel economy loss on 91 octane
     unleaded 147
   lead content of 24
   octane quality 23
   physical properties of 46
   powered vehicles, emission factors for .53, 54
   quality trends 23
   retail price of 166, 167
   savings, projected potential 1985 .... 35, 36
   unleaded 113
   volatility 24
Geographic consideration 201
Great Britain in 1934, vehicle road
   tax in 162, 163
Great Britain, new registration of
   taxis in 164, 165
Great Britain, running costs in
   1933 in 160, 163
Growth in vehicle miles traveled 25, 26
Growth, moderating gasoline demand 29

H

Heat of combustion
   calculated and measured 2, 4
   equation for 2
   as a function of hydrogen content 2, 3
   as a function of molecular weight 1, 3
Heavy duty engine 185, 198
High speed diesel engine 159
History of major automotive products,
   commercialization 194, 196
History of modern OHV engines,
   commercialization 195, 196
Honda Civic 88, 189, 192
Honda compound vortex-controlled
   combustion system (CVCC) 86
   conversion of 350-CID Chevrolet
     Impala 89, 90
   emissions 88, 89, 90
   engine 86, 87
Honda engine, fuel economy of 89
Hornets, field test of Du Pont muffler lead
   traps on 258 CID 131
Hydrocarbon
   –air combustion rates 79
   emissions 61, 80, 84, 91, 93, 95
     as a function of air–fuel equivalence
       ratio 43, 45
     with gaseous fuels 53

Hydrocarbon (continued)
   reduction in 43
   response to mid-range and back-end
     volatility 62, 65
   liquid fuels 1
Hydrocarbons
   from coke, energy recovery of 6, 8
   from coke, weight recovery of 6, 8
   EPA air quality standards for 100
   EPA automotive emission standards for ... 100
   to gaseous fuel, liquid 69
   from shale and coal 7, 11
   related to the hydrogen content,
     properties of 2
   shape selective cracking of C_6 73
   viscosity of C_{26} saturated 2, 4
Hydrogen from coke, production of pure ... 6
Hydrogen content
   of fossil fuels 2, 5
   heat of combustion as a function of 2, 3
   materials, fuels from lower 6
   octane quality and total 12
   properties of hydrocarbons related
     to the 2
   of the raw material, usable 6
Hydrogenation
   coal 9
   energy recovery from bituminous
     coal 14, 15
   properties of products from coal 12, 13
   units 14
   weight recovery from bituminous
     coal 14, 15
Hypothetical future new car sales mixes. 30, 31

I

Ignition engines, reciprocating spark 82
Ignition engines, torch 86
Indirect chamber engine 160
Injection, fuel 82, 162
Interim standards, 1975 California 140
Interim standards, 1975 U.S. 139
Investment, capital 194

J

Jeeps with stratified engine conversions   84

L

Lead
   balance of muffler lead trap 128, 129
   content of gasoline 24
   emissions, measurement of 131, 134
   emission rate 129, 130, 132, 133
   on gasoline demand, added effect of 33
   particles from TECS Pinto, size
     distribution of emitted 130
   tolerant emission control systems 126
   trap, muffler (see Muffler lead trap)
Limits, emission 29
Liquid fuels from coal 16
Liquid fuels, hydrocarbon 1
Liquid hydrocarbons to gaseous fuel 69
Liquified petroleum gases (LPG) 47
LPG (liquefied petroleum gases) 47
   fuel system 48, 49

Low emissions combustion engines 78
Low mileage emission levels—Ford
    PROCO conversions 85

M

Manifold thermal reactor, exhaust 114
Market, candidate alternate
    powerplants by 173
Market penetration, range of
    expected 198, 199
Market share projections 197
Maximum carbon monoxide values
    in Chicago 105, 106
Mazda RX2 ....189, 192
Measured heats of combustion,
    calculated and 2, 4
Mercedes Benz chamber 160, 161
Methane, physical properties of 46
Metropolitan centers, daily area traffic
    volumes in 110, 111
Mid-range and back-end volatility,
    emissions response to 62, 63, 65
Mileage emissions levels—Ford PROCO
    conversions, low 85
Miles, reduced vehicle 30
Miles traveled, growth in vehicle 25, 26
Modifications
    to carburetion system 63
    combustion 79
    emissions with only engine 125
    for gaseous fuels, retrofit 48
Modified combustion engines 79
Modified high compression ratio
    Chevrolet 125, 126
Modified rollback 103
Molecular weight, heat of combustion as
    a function of 1, 3
Monitoring program (CAMP), EPA's
    continuous air 102
Motor fuels, gaseous 43, 47
Motor vehicle emission improvements
    with gaseous fuels 55, 56
Motor vehicles, low emissions combus-
    tion engines for 78
Muffler lead trap 114
    Du Pont 126, 131, 133
    lead balance of 128, 129
    on 1970 Chevrolet, Du Pont 127, 128
    on 258 CID 1971–72 Hornets 131

N

Naphtha, Kuwait 73, 75
Natural gas 47
Nitric oxide formation 79, 80, 81
Nitrogen dioxide, EPA air quality
    standards for 100
Nitrogen oxides (NO$_x$)
    emissions 79, 80, 84, 89, 91, 93, 95
    conventional and divided
        combustion chamber 91, 92
    as a function of air–fuel
        equivalence ratio 44, 45
    with gaseous fuels 53
    standard 113
    EPA automotive emission
        standards for 100

Nitrogen oxides (NO$_x$) *(continued)*
    response to mid-range and back-end
        volatility 63, 65

O

Octane
    number 12, 23, 73
    quality, gasoline 12, 23
    rating for C$_5$–400°F reformate 71, 72
    rating *vs.* conversion level 73, 75
    rating for Kuwait naphtha 73, 75
    requirement reduction 149, 150
    requirements, research 148, 149
    unleaded gasoline, 91 147
OHV engines 195, 196
Oil, retorted 9
Oil, shale 9
    organic material in 9, 10
Open-chamber stratified charge engines ... 82
Organic material in oil shale 9, 10
Oxidative regeneration, catalyst life and
    stability toward 76

P

Packaging, passenger car engine 186
Parameters, selection 178, 179, 185, 186, 187
Particles from TECS Pinto, emitted lead 130
Passenger car engines 177, 197
    packaging 186
Passenger car trends 21, 22
Performance
    on 1970 Chevrolets, muffler
        lead trap 127, 128
    of diesel and gasoline engines 169, 171
    of gaseous-fueled vehicles 50
    with TECS 129
    between turbocharged diesel and gaso-
        line engines, test bed 169, 170
Petroleum gases (LPG), liquified 47
Petroleum refining 2
Photochemical oxidants, EPA air quality
    standards for 100
Pinto
    emitted lead particles from TECS 130
    exhaust emission levels of
        TECS-equipped 120, 122
    exhaust gas recirculation system
        for a 1.6 l 120, 121
    TECS III on a 1.6 l 118, 119
    thermal reactor for a 1.6 l 117, 119
    with and without TECS III, fuel
        economy of 1971 ....123, 124
    with and without TECS III, total lead
        emission rate from 1971 ....129, 130
Piston engine, design flexibility
    of the 191, 192
Plant, gasoline–diesel 16
Pollutant emissions, effect of gas-powered
    vehicles on 55, 56
Pollution control, attainment of air 101, 104
Power as a function of air–fuel
    equivalence ratio 44, 49
Power loss with gaseous fuels 44
Power unit, saloon car 159
Powered saloon cars, diesel- 162

Powerplants by market, candidate
alternate 173
Prechamber stratified charge
engines 85, 86, 87
Predicting emission standards 105
Pre-engine converter 69
Price of diesel and gasoline, retail 166, 167
Processes, continuous combustion 94
Processing schemes, synthetic 9
PROCO, (Ford-programed combustion
system) 82, 83, 85
Product distribution at 900°F 73, 74
Production, diesel car 166, 167
Production of pure hydrogen from coke   6
Products from coal hydrogenation, of 12, 13
Products, current specification 2, 5
Projected carbon monoxide
concentration 103, 106
Projected potential 1985 gasoline
savings 35, 36
Projections, future engine
use 197, 198, 199, 200
Projections, gasoline demand 24
effect of efficiency improvement on 33, 34
effect of emission standards on 28, 31
effect of new car sales mix on 32, 34
hypothetical future new car sales
mixes for 30, 31
Projections, market share 197
Propane, physical properties of 46

Q

Quality and total hydrogen content,
octane 12
Quality trend analysis, air 103

R

Rankine cycle engines 94
Reactor, catalytic 69, 70
Reactor for a 350 CID
Chevrolet 116, 119, 120, 121
Reactor, thermal (see Thermal reactor)
Reciprocating spark ignition engines 82
Recirculation (EGR), exhaust
gas 114, 120, 121
Recovery
from coal hydrogenation, energy 14, 15
from coal hydrogenation, weight 14, 15
of hydrocarbons from coke, energy 6, 8
of hydrocarbons from coke, weight ... 6, 8
of hydrocarbons from shale and
coal, energy 7, 11
of hydrocarbons from shale and
coal, weight 7, 11
Reduction, compression ratio 113, 136, 138
Reduction, octane requirement 149, 150
Reduction in hydrocarbon emission 43
Reformate, C₅–400°F 71, 72
Registrations of taxis in Great
Britain, new 164, 165
Regulations on gasoline demand, impact
of emissions 19
Refining, petroleum 2
Regeneration, oxidative 76
Research octane requirements 148, 149
Requirement reduction, octane 149, 150

Requirements, research octane 148, 149
Requirements, social 177, 179
Research octane number (RON) 23
Resources, total energy 1
Response to mid-range and back-end
volatility, emissions 62, 63, 65
Ricardo Comet chamber 160, 161
Road speed fuel consumption,
constant 164, 165, 171
Road tax in Great Britain in 1934,
vehicle 162, 163
Retorted oil 9
Retrofit modifications for gaseous fuels   48
Rollback approach, simple 102
Rollback, modified 103
RON (research octane number) 23
Rotary (Wankel) engine 174, 178
Running costs in 1933 in
Great Britain 160, 163

S

Safety of gaseous fuels 51
Sales mix on gasoline demand projections,
effect of new car 32, 34
Saloon car power unit 159
Saloon cars, diesel powered 162
San Francisco field test of Du Pont
muffler lead traps 131
Saturated hydrocarbons, viscosity of C₂₆   2, 4
Savings, gasoline 35, 36
Selection parameters 178, 179, 185, 186, 187
Schemes, synthetic processing 9
Selective cracking of C₆ hydrocarbons,
shape 73
Shale, energy recovery of hydrocarbons
from 7, 11
Shale oil 9, 10
Shale, weight recovery of hydrocarbons
from 7, 11
Shape selective cracking of C₆
hydrocarbons 73
Single-cylinder divided combustion
chamber engine 93
Size distribution of lead particles from
TECS Pinto 130
Small engines 186, 198
Smaller cars, shift to 30
Social requirements 177, 179
Source of energy, unaltered carbon as a   6
Spark ignition engines, reciprocating 82
Specification products, current 2, 5
Speed fuel consumption, constant
road 164, 165, 171
Staged combustion engine,
divided-chamber 90, 93
Staging of combustion process, timed 80
Standard high compression ratio
Chevrolet, fuel economy with 125, 126
Standards
1975 California interim 140
1975 U.S. interim 139
1976 140
1977 140
air quality 102, 110, 111
carbon monoxide 112, 115
emissions (see Emissions standards)

Standards, EPA 100
Steady state emissions from Honda CVCC
   conversion of 350-CID Chevrolet 90
Stirling cycle, engine 94, 175, 182
Storage of gaseous motor fuels 47
Stratification, charge 85
Stratified charge engine 93, 175, 183
   open-chamber 82
   prechamber 85
   Wankel 185
Studies, air quality 112
Sulfur dioxide emissions with gaseous
   fuels 57
Supply of gaseous fuel 50
Synthesis processes, fuel 14
Synthetic processing schemes 9
System, modifications to carburetion 63

T

Tax in Great Britain in 1934,
   vehicle road 162, 163
Taxis in Great Britain, new
   registrations of 164, 165
TCCS (Texaco-controlled combustion
   system) 82, 83
TECS (Du Pont total emission control
   system) 114, 115
TECS III
   design of 116
   emission control with 120
   emission levels of vehicles with 120, 123
   fuel economy of Chevrolets with 123, 124
   on a 1.6 1 Pinto 118, 119
   1971 Pinto with and
     without 123, 124, 129, 130
   vehicles 117
TECS IV
   design of 118
   emission control with 120
   emission levels of vehicles 120, 123
   fuel economy of Chevrolets with 123, 124
TECS-equipped Pinto 120, 122, 130
TECS, performance with 129
Test bed performance 169, 170
Test of Du Pont muffler lead traps 131
Test fuels 62, 68, 71
Tests, engine emissions 93
Texaco-controlled combustion system
   (TCCS) 82, 83
Thermal efficiency of fuel synthesis
   processes 14
Thermal emission control systems on fuel
   economy, effects of 143, 145, 146,
                  151, 152
Thermal emission control systems,
   Du Pont 142
Thermal reactor
   control systems, effect of 141
   -EGR emission control systems 144, 145
   exhaust manifold 114
   for a 350 CID Chevrolet 116, 119, 120, 121
   for a 1.6 1 Pinto 117, 119
Timed combustion chamber fuel injection 82
Timed staging of combustion process 80

Torch ignition engine 86
Total emission control system (TECS),
   Du Pont 114, 115
Traffic activity 103, 104
Traffic volume and allowable carbon
   monoxide emission rate 109
Traffic volumes in metropolitan
   centers 110, 111
Traps, muffler lead (see Muffler lead
   traps)
Trend analysis, air quality 103
Trends
   on gasoline demand, impact of
     automotive 19
   gasoline quality 23
   new car equipment 21, 22
   passenger car 21, 22
   in six cities, carbon monoxide 107, 108
Turbine engine 94, 174, 181
Turbocharged diesel and gasoline
   engines 169, 170
Types, engine 174

U

Unaltered carbon as a source of energy   6
U.S. interim standards, 1975 139
Unit, saloon car power 159
Units, hydrogenation 14
Unleaded gasoline 113, 147

V

Vehicle (see also Automobile and Car)
   acceleration with a gasoline and
     diesel engine 160, 163
   carbon monoxide emission rate 105, 106
   carbon monoxide emission
     standards 107, 108, 110, 111
   efficiency, improved 32
   emission improvements with gaseous
     fuels 55, 56
   fuel economy 113
   miles, reduced 30
   miles traveled, growth in 25, 26
   road tax in Great Britain in 1934 ...162, 163
Vehicles
   fuel economy for current production.... 85
   gasoline and gas powered 53, 54, 55, 56
   low emissions combustion engines for
     motor 78
   performance of gaseous-fueled 50
   with TECS III and TECS IV....117, 120, 123
Viscosity of C_{26} saturated hydrocarbons 2, 4
Volatility
   change in carbon dioxide emissions,
     effect of 66
   characteristics, test fuels 62
   emissions response to mid-range and
     back-end 62, 63, 65
   fuel 60
   gasoline 24
   relationship of fuel economy to 68
Volume, catalytic reactor 70
Vortex-controlled combustion system
   (CVCC), Honda compound 86

W

Wankel engine 174, 178, 185

Weight, heat of combustion as a function
of molecular 1, 3

Weight recovery from bituminous coal
hydrogenation 14, 15

Weight recovery of hydrocarbons from
coke 6, 8

Weight recovery of hydrocarbons from
shale and coal 7, 11

Z

Zeolite catalyst   71